SOLITON MANAGEMENT IN PERIODIC SYSTEMS

SOLITON MANAGEMENT IN PERIODIC SYSTEMS

Boris A. Malomed
Tel Aviv University
Israel

 Springer

Boris A. Malomed
Tel Aviv University, Israel

Soliton Management in Periodic Systems

Consulting Editor: D. R. Vij

ISBN 0-387-29334-5 (e-book)
ISBN: 978-1-4419-3817-6
ISBN: 978-0-387-29334-9 (e-book)
Printed on acid-free paper.

9 8 7 6 5 4 3 2 1

springeronline.com

Contents

Preface

This books makes an attempt to provide systematic description of recently accumulated results that shed new light on the well-known object – solitons, i.e., self-supporting solitary waves in nonlinear media. Traditionally, solitons are studied theoretically (in an analytical and/or numerical form) as one-, two-, or three-dimensional solutions of nonlinear partial differential equations, and experimentally – as pulses or beams in uniform media. Propagation of solitons in inhomogeneous media was considered too (chiefly, in a theoretical form), and a general conclusion (which could be easily expected) was that the soliton would suffer gradual decay in the case of weak inhomogeneity, and faster destruction in strongly inhomogeneous systems.

However, it was recently found, in sundry physical and mathematical settings, that a completely different, and much less obvious, situation is possible too – a soliton may remain a truly robust and intrinsically coherent object traveling long distances in periodic heterogeneous media, composed of layers with very different properties. A well-known example is *dispersion management* in fiber-optic telecommunications, i.e., the situation when a long fiber link consists of periodically alternating segments of fibers with opposite signs of the group-velocity dispersion. Such a structure of the link is necessary, as the dispersion must be compensated on average, which is provided by the alternation of negative- and positive-dispersion segments. In this case, a simple result is that localized pulses of light feature periodic internal pulsations but remain stable on average (do not demonstrate systematic degradation) in the absence of nonlinearity. A really nontrivial result is that optical solitons, i.e., *nonlinear* pulses of light, may also remain extremely stable propagating in such a periodically heterogeneous system. Moreover, under certain conditions, (quasi-) solitons may be robust even in a *random* dispersion-managed system, with randomly varying lengths of the dispersion-compensated cells (each cell is a pair of fiber segments with opposite signs of the dispersion).

While the dispersion management provides for the best known example of stability of solitons under "periodic management", examples of robust oscillating solitons in periodic heterogeneous systems were also found and investigated in some detail in a number of other settings. Essentially, they all belong to two areas – nonlinear optics and Bose-Einstein condensation (being altogether different physically, these fields have a lot in common as concerns their theoretical description). It should be said that the action of the periodic heterogeneity on a soliton may be realized in two different ways – as motion of the soliton through the inhomogeneous medium, or as strong periodic variation of system's parameter(s) in time, while the soliton does not move at all.

A very interesting example of the latter situation is the so-called *Feshbach-resonance management*, when the sign of the effective nonlinearity in a Bose-Einstein condensate periodically changes between self-attraction and self-repulsion. In the latter case, nontrivial examples of stable solitons have also been predicted.

The book aims to summarize results obtained in this field. In fact, a vast majority of results still have the form of theoretical predictions, as systematic experimental study of stability of solitons in periodic heterogeneous systems have only been performed in the context of the dispersion management in fiber optics. For this reason, the material collected in the book has a strong theoretical bias. A hope is that collecting the theoretical predictions in a systematic form may suggest directions for experimental investigation of solitons under the "periodic management". In particular, creation of solitons in Bose-Einstein condensates subjected to the Feshbach-resonance management, possibly in combination with spatially periodic potentials, provided by the so-called optical lattices, seems to be quite feasible in the real experiment, which would be especially interesting in two- and three-dimensional settings (creation of a three-dimensional soliton in a real experiment has never been reported in any field of physics, despite various theoretical predictions of this possibility).

As concerns theoretical results, virtually all of them are not rigorous ones, for an obvious reason – it is very difficult to rigorously prove the existence of stable oscillating localized solutions in models based on nonlinear partial differential equations with periodically varying coefficients, which provide for the theoretical description of the systems with periodic management. Therefore, theoretical results are either purely numerical ones, or, sometimes, they are known in a (semi-) analytical form, which is based (most frequently) on the variational approximation. Nevertheless, despite the lack of the rigorous theory, there is a possibility to summarize the results in a systematic and sufficiently consistent form. An attempt of that is done in this book. It should be said that the presentation of material in the book has a rather subjective character (which is, probably, inevitable in a book on such a topic), as emphasis is made on those issues and aspects which seem specially interesting or significant from the viewpoint of the author.

The subject of the periodic management of solitons is far from being completed. Not only the experimental results are very scarce, as said above, but also theoretical analysis (even in a non-rigorous form) of many important problems should be further advanced. However, although the field is in the state of development, a coherent description of its current status is quite possible.

Three distinct parts can be identified in the book. The first chapter (Introduction), which is, as a matter of fact, a separate part by itself, gives a possibly general overview of solitons, with an intention to briefly outline the most important theoretical models and results obtained in them, as well as most significant experimental achievements. Since the length of the introduction is limited, the outline was focused on models and settings related to the realms of nonlinear optics and Bose-Einstein condensation, as the concepts and techniques of the periodic soliton managements have been developed in these areas. The introduction also includes a brief description of the subject and particular objectives of the book. Then, two technical parts (one includes chapters 2 – 6, and the other chapters 7 – 10) report results, respectively, for one-dimensional and multidimensional solitons. Such separation is natural, as methods used for the study of one-dimensional settings, and the respective results, are very different from those which are relevant to multidimensional problems (nevertheless, chapter 10 includes some results for a one-dimensional situation too, which are closely related to the basic two-dimensional problem which is considered in that chapter).

Writing this book would not be possible without valuable collaborations and discussions with a large number of colleagues. It is my great pleasure to express the gratitude to F. Kh. Abdullaev, J. Atai, B. B. Baizakov, Y. B. Band, A. Berntson, J. G. Caputo, A. R. Champneys, P. Y. P. Chen, P. L. Chu, D. J. Frantzeskakis, B. V. Gisin, D. J. Kaup, P. G. Kevrekidis, Y. S. Kivshar, R. A. Kraenkel, T. Lakoba, U. Mahlab, D. Mihalache, V. Pérez-García, M. Salerno, M. Segev, N. Smyth, L. Torner, M. Trippenbach, F. Wise, and J. Yang. Special thanks are due to younger collaborators (some of them were my students or postdoc associates), including R. Driben, A. Gubeskys, M. Gutin, A. Kaplan, M. Matuszewski, T. Mayteevarunyoom, M. I. Merhasin, G. Theocharis, and I. Towers.

The work on particular projects that have generated essential results included in this book was supported, in various forms and parts, by grants No. 1999459 from the Binational (US-Israel) Science Foundation, and No. 8006/03 from the Israel Science Foundation. At a smaller scale, support was also provided by the European Office of Research and Development of the US Air Force, and Research Authority of the Tel Aviv University.

List of acronyms used in the text:

1D, 2D, 3D - one-dimensional, two-dimensional, three-dimensional

AWG - antiwaveguide

BEC - Bose-Einstein condensation/condensate

BG - Bragg grating

CW - continuous-wave (solution)

DM - dispersion management

DS - dark soliton

FF - fundamental-frequency (wave)

FP - fixed point

FR - Feshbach resonance

FRM - Feshbach-resonance management

FWHM - full width at half-maximum (of an optical pulse)

FWM - four-wave mixing

GPE - Gross-Pitaevskii equation

GS - gap soliton

GVD - group-velocity dispersion

GVM - group-velocity mismatch

HS - hot spot (a local perturbation switching a spatial soliton)

ISI - inter-symbol interference

IST - inverse-scattering transform

KdV - Korteweg - de Vries (equation)

ME - Mathieu equation

NLM - nonlinearity management

NLS - nonlinear Schrödinger (equation or soliton)'

ODE - ordinary differential equation

OL - optical lattice

PAD - path-average dispersion

PCF - photonic-crystal fiber

PDE - partial differential equation

PR - parametric resonance

QPM - quasi-phase-matching

RI - refractive index

RZ - return-to-zero (signal)

SH - second harmonic

SHG - second-harmonic generation

SPM - self-phase modulation

SSM - split-step model

STS - spatiotemporal soliton

TF - Thomas-Fermi (approximation)

TS - Townes soliton

VA - variational approximation

WDM - wavelength-division multiplexing

WG - waveguide (when referred to in the context of the waveguding-antiwaveguiding model)

XPM - cross-phase modulation

Chapter 1

Introduction

1.1 An overview of the concept of solitons

The concept of *solitons* (solitary waves) plays a profoundly important role in modern physics and applied mathematics, extending beyond the bounds of these disciplines. It was introduced in 1965 by Zabusky and Kruskal who numerically simulated collisions between solitary waves (pulses) in the Korteweg - de Vries (KdV) equation, and discovered that these pulses not only are stable in isolation, but also completely recover their shapes after collisions [175]; this observation was an incentive which had soon led to the discovery of the *inverse scattering transform* (IST) and the very concept of integrable nonlinear partial differential equations (PDEs) [72]. The next principally important step in this direction was made by Zakharov and Shabat, who had demonstrated that the integrability is not a peculiarity specific to a single (KdV) equation, but is also featured by another equation which finds very important applications in physics, *viz.*, the nonlinear Schrödinger (NLS) equation [177]. Integrability of the sine-Gordon equation, which was actually known, in terms of the Bäcklund transformation, since the 19th century, was also naturally incorporated into the IST technique (the sine-Gordon equation finds its most important physical realization in superconductivity, as a dynamical model of a long *Josephson junction*, i.e., a thin layer of an insulator sandwiched between two bulk superconductors [170]). Further development of the studies in this field has produced a body of results which have become a classical contribution to several core areas of physics and mathematics. The IST technique and results produced by it were summarized in several well-known books written by the very same people who had produced these results [176, 11, 133].

Parallel to the theoretical developments, great progress has been achieved in experimental studies of solitons. The very first published report of observation of a soliton is due to John Scott Russell, who spotted a stable localized elevation running on the surface of water in a canal in Edinburgh, and pursued it on horseback. In retrospective, the most astonishing feature of this report, published in 1844 [149], is the very fact that J. S. Russell was able to instantaneously understand the significance of the phenomenon.

1.1.1 Optical solitons

Qualitative consideration

In the modern experimental and theoretical studies of solitons, the most significant progress has been achieved in optics and, most recently, in Bose-Einstein condensates (BECs). A milestone achievement was the creation of bright temporal solitons in non-linear optical fibers in 1980 [127], after this possibility had been predicted seven years earlier [79]. In the realm of nonlinear optics, this was followed by the creation of dark solitons in fibers [60, 98, 172], bright spatial solitons in planar nonlinear waveguides [118, 18], and *gap solitons* (GSs) in fiber Bragg gratings [57]. In all these cases, the soliton is supported by interplay between the chromatic dispersion (in the temporal domain) or diffraction (for spatial solitons) of the electromagnetic wave and cubic self-focusing nonlinearity, induced by the Kerr effect. The latter may be realized as an effective positive correction, $\Delta n(I)$, to the local refractive index (RI) of the material medium, which is proportional to the local intensity, I, of that very electromagnetic wave on which the RI acts, i.e., $\Delta n(I) = n_2 I$ with a positive coefficient n_2. Besides the *self-focusing* sign of the Kerr effect ($\Delta n(I) > 0$), its essential property in normal optical materials is the instantaneous character (no temporal delay between $\Delta n(I)$ and I). In view of the fundamental importance of the temporal and spatial optical solitons supported by this mechanism, it is relevant to present a short quantitative explanation for it here.

In the course of the propagation in the nonlinear medium, the light pulse accumulates a phase shift that, through the correction $n_2 I$ to the RI, mimics the temporal shape of the pulse, $I = I(t)$. To understand this feature in a more accurate form, one may start from the normalized wave equation for the electric field E,

$$E_{zz} + E_{xx} + E_{yy} - \left(n^2 E\right)_{tt} = 0, \tag{1.1}$$

where the subscripts stand for the partial derivative, z is the propagation distance, x and y are transverse coordinates, t is time, and n is the above-mentioned RI (detailed derivation of the wave equation can be found, e.g., in book [15]). A solution to Eq. (1.1) for a one-dimensional wave, which must be a real function, is looked for as

$$E(z,t) = u(z)e^{ik_0 z - i\omega_0 t} + u^*(z)e^{-ik_0 z + i\omega_0 t}, \tag{1.2}$$

where $\exp\left(ik_0 z - i\omega_0 t\right)$ represents a rapidly oscillating carrier wave, the asterisk stands for the complex conjugation, and $u(z,t)$ is a slowly varying complex local amplitude. Substituting this in Eq. (1.1), in the lowest approximation one obtains the dispersion relation between the propagation constant (wave number) k and frequency ω, $k_0^2 = \left(n_0 \omega_0\right)^2$, with n_0 the RI in the linear approximation. The next-order approximation, which takes into regard the above correction to the RI, $n = n_0 + n_2 I$, yields an evolution equation for the amplitude,

$$i\frac{du}{dz} + \frac{n_0 n_2}{k_0}\omega_0^2 I u = 0. \tag{1.3}$$

Actually, this equation is a nonlinear one, as the intensity is tantamount to the squared amplitude, $I = |u|^2$. A solution to Eq. (1.3) is simply $\Delta\phi = \left(n_0 n_2\right)\left(\omega_0^2/k_0\right)Iz$, where $\Delta\phi$ is a nonlinear contribution to the wave's phase (the accumulation of the non-linear phase is usually called self-phase modulation, SPM). The corresponding SPM-induced frequency shift being $\Delta\omega = -\partial\Delta\phi/\partial t$, one obtains an expression for it,

$$\Delta\omega = -n_0 n_2 \frac{\omega_0^2}{k_0} \frac{dI}{dt} z. \tag{1.4}$$

It follows from Eq. (1.4) that the lower-frequency components of the pulse, with $\Delta\omega < 0$, develop near its front, where $dI/dt > 0$ (the intensity grows with time), while higher frequencies, with $\Delta\omega > 0$, develop close to the rear of the pulse, where $dI/dt < 0$.

On the other hand, the dielectric response of the material medium is not strictly instantaneous, featuring a finite temporal delay. This implies that the linear part, $\epsilon \equiv n_0^2$, of the multiplier n^2 in the wave equation (1.1) (the dynamic dielectric permeability) is, as a matter of fact, a linear operator, rather than simply a multiplier. The accordingly modified form of the linear term $(\epsilon E)_{tt}$ in Eq. (1.1) becomes $\left(\int_0^\infty \epsilon(\tau)E(t-\tau)d\tau\right)_{tt}$, where τ is the delay time. Finally, approximating this nonlocal-in-time expression by a quasi-local expansion, $\epsilon_0 E_{tt} + \epsilon_2 E_{tttt} + ...$, which is justified when the actual delay in the dielectric response is very small, gives rise to second- and higher-order group-velocity-dispersion (GVD), alias chromatic-dispersion, terms in the eventual propagation equation, which can be translated into the corresponding linear dispersion relation, $k = k(\omega)$ [15].

In particular, the normal (positive) GVD (which means that waves with a higher frequency have a smaller group velocity, as expressed by the condition that the second-order-dispersion coefficient is positive, $\beta_2 \equiv d^2k/d\omega^2 > 0$) reinforces the above (nonlinearity-induced) trend to the temporal separation between the low- and high-frequency components of the pulse, contributing to its rapid spread. On the contrary, anomalous (negative) GVD ($\beta_2 < 0$), which also occurs in real materials, may *compensate* the nonlinearity-induced spreading. With the magnitudes of the dispersion and intensity properly matched, the balance may be perfect, giving rise to very robust pulses, i.e., solitons.

Nonlinear Schrödinger equation and solitons

Putting all the above ingredients together, and assuming that the amplitude u in Eq. (1.2) is a slowly varying function of z and "reduced time", $\tau \equiv t - k'_\omega z$ (here and below, the value of the derivative k'_ω is calculated at the carrier-wave's frequency, $\omega = \omega_0$), one arrives at the *nonlinear Schrödinger* (NLS) equation which governs the evolution of $u(z, \tau)$,

$$iu_z - \frac{1}{2}\beta u_{\tau\tau} + \gamma|u|^2 u = 0, \tag{1.5}$$

where β replaces β_2 (the replacement will not lead to confusion, as higher-order dispersion, which is different from β_2, is not dealt with below), and $\gamma \equiv n_2\sqrt{\epsilon_0}\omega_0^2/k_0$. The introduction of τ instead of t is necessary to eliminate a term with the first derivative in t (the group-velocity term), thus casting the NLS equation in the simplest possible form, namely, the one given by Eq. (1.5).

Below, a number of models will be considered that may be viewed as various generalizations of the NLS equation (1.5) – two-component systems, equations with different nonlinearities, multidimensional systems, etc. A very recent succinct review of equations of the NLS type can be found in article [105].

An elementary property of the NLS equation is its *Galilean invariance*: any given solution $u(z, \tau)$ automatically generates a family of moving solutions by means of the *Galilean boost* that depends on an arbitrary real parameter c (it is an inverse-velocity shift, relative to the inverse group velocity, k'_ω, of the carrier wave):

$$u(z, t; c) = u\left(z, \tau - cz\right) \exp\left(\frac{ic^2}{2\beta}z - \frac{ic}{\beta}\tau\right). \tag{1.6}$$

Another simple property of Eq. (1.5) is the *modulational instability* of CW (continuous-wave) solutions, $u_{\mathrm{CW}} = A_0 \exp\left(i\gamma A_0^2 z\right)$ with an arbitrary amplitude A_0: although the CW solution does not contain the GVD coefficient β, it is stable in the case of $\beta\gamma < 0$, and unstable (against τ-dependent perturbations) in the opposite case.

The NLS equation has natural Lagrangian and Hamiltonian representations. The former one will be considered below (see Eq. (2.7)), while the latter takes the form

$$iu_z = \frac{\delta H}{\delta u^*}, \tag{1.7}$$

where $\delta/\delta u^*$ is the functional derivative, the asterisk stands for the complex conjugation, and the Hamiltonian,

$$H = -\frac{1}{2}\int_{-\infty}^{+\infty}\left(\beta\left|u_\tau\right|^2 + \gamma|u|^4\right) d\tau, \tag{1.8}$$

is considered as a functional of two formally independent arguments, $u(\tau)$ and $(u(\tau))^*$. The Hamiltonian is a dynamical invariant of Eq. (1.5), i.e., $dH/dz = 0$. Two other straightforward dynamical invariants of the NLS equation are energy E, alias norm of the solution (in the context of fiber optics, the energy is different from the Hamiltonian), and momentum P,

$$E \equiv \frac{1}{2}\int_{-\infty}^{+\infty}|u(\tau)|^2 d\tau, \tag{1.9}$$

$$P \equiv i\int_{-\infty}^{+\infty} uu_\tau^* d\tau. \tag{1.10}$$

Due to the fact that the NLS equation is exactly integrable by means of the IST, it has an infinite set of higher-order dynamical invariants, in addition to E, P, and H [176]. In particular, the first two higher-order invariants are

$$I_4 = \frac{1}{2}\int_{-\infty}^{+\infty}\left(-\beta uu_{\tau\tau\tau}^* + 3\gamma|u|^2 uu_\tau^*\right) d\tau, \tag{1.11}$$

$$I_5 = \frac{1}{4}\int_{-\infty}^{+\infty}\left[\beta^2\left|u_{\tau\tau}\right|^2 + 2\gamma^2|u|^6 + \gamma\beta\left((|u|^2)_\tau\right)^2 + 6\gamma\beta\left|u_\tau\right|^2 u^2\right] d\tau. \tag{1.12}$$

(the subscripts 4 and 5 imply that they follow the first three elementary dynamical invariants, E, P, and H). These higher-order invariants do not have a straightforward physical interpretation, and are seldom used in applications. Nevertheless, an example of a physical application of the invariants (1.11) and (1.12) will be presented in this book, when analyzing splitting of higher - order solitons in the model based on Eq. (5.5), see subsection 5.2.3.

In the case of the anomalous GVD, $\beta < 0$ (it is assumed that γ is positive), i.e., when the CW solutions are unstable, a commonly known family of soliton solutions to Eq. (1.5) is

$$u_{\mathrm{sol}}(z,\tau) = \frac{\eta}{\sqrt{\gamma}}\operatorname{sech}\left(\eta\left(\frac{\tau}{\sqrt{|\beta|}} - cz\right)\right)\exp\left(i\left[\frac{c\tau}{\sqrt{|\beta|}} + \frac{1}{2}\left(\eta^2 - c^2\right)z\right]\right),$$
(1.13)

where η and c are arbitrary real parameters, that determine the soliton's amplitude and the above-mentioned inverse-velocity shift. The function sech (hyperbolic secant) in this solution provides for the localization of the soliton. In the experiment, the temporal soliton is observed as a localized pulse running along the fiber with the velocity $V = 1/\left(k'_\omega + c\sqrt{|\beta|}\right)$. The entire soliton family (1.13) is stable against small perturbations.

The application of the IST yields exact solutions of the NLS equation more complex than the fundamental soliton (1.13). In particular, the initial condition (in the case of $\beta < 0$)

$$u_0(\tau) = n\frac{\eta}{\sqrt{\gamma}}\operatorname{sech}\left(\frac{\eta}{\sqrt{|\beta|}}\tau\right)$$
(1.14)

with integer n and arbitrary η, that generates the fundamental soliton for $n = 1$, gives rise to higher-order "n-solitons" for $n \geq 2$ [154]. Analytical expressions for these solitons with $n \geq 3$ are cumbersome. A relatively simple analytical solution describes the 2-soliton,

$$u_{\mathrm{2sol}} = \frac{4\eta}{\sqrt{\gamma}}\frac{\cosh\left(3\eta\tau/\sqrt{|\beta|}\right) + 3\exp\left(4i\eta^2 z\right)\cosh\left(3\eta\tau/\sqrt{|\beta|}\right)}{\cosh\left(4\eta\tau/\sqrt{|\beta|}\right) + 4\cosh\left(2\eta\tau/\sqrt{|\beta|}\right) + 3\cos\left(4\eta^2 z\right)}\exp\left(\frac{i}{2}\eta^2 z\right).$$
(1.15)

As seen from this expression, the shape of the 2-soliton, i.e., the distribution of the power in the soliton, $|u(z,\tau)|^2$, oscillates in z with the period

$$z_{\mathrm{sol}} = \frac{\pi}{2\eta^2},$$
(1.16)

which is called the *soliton period*. It can be demonstrated that all the exact n-soliton solutions generated by the initial condition (1.14) with $N \geq 2$ oscillate with exactly the same period (1.16), irrespective of the integer value of n. In fact, z_{sol} is also an estimate for the propagation distance which is necessary for formation (self-trapping) of the fundamental soliton from an initial pulse of a generic form.

As well as the fundamental soliton (1.13), the 2-soliton (1.15) remains single-humped at any z (i.e., $|u(z, \tau)|^2$ always has a single maximum as a function of τ). However, the 3-soliton solution periodically splits into a double-humped structure and recombines into a sharp single-peak one, see Fig. 5.4 in book [15].

In terms of the IST, the 2-soliton (1.15) may be regarded as a nonlinear bound state of two fundamental solitons, with the amplitudes

$$\eta_1^{(n=2)} = 3\eta, \ \eta_2^{(n=2)} = \eta. \tag{1.17}$$

Similarly, the 3-soliton is a bound state of three fundamental solitons, with

$$\eta_1^{(n=3)} = 5\eta, \ \eta_2^{(n=3)} = 3\eta, \ \eta_3^{(n=3)} = \eta. \tag{1.18}$$

Note that the energy (1.9) of the n-soliton (1.14) is

$$E_n = \frac{2\sqrt{|\beta|}}{\gamma} n^2 \eta. \tag{1.19}$$

As follows from the above and Eq. (1.19), for $n = 2$ and $n = 3$ (actually, for any n) the energy of the n-soliton is exactly equal to the sum of energies of the constituent fundamental solitons, if they are separated from each other. To understand if the bound state is stable against splitting into the separate fundamental solitons, one can identify its *binding potential*, as a difference between the value of the Hamiltonian (1.8) for the n-soliton, which is

$$H_n = \frac{\sqrt{|\beta|}}{3\gamma} \eta^3 n^2 \left(2n^2 - 1\right), \tag{1.20}$$

and the sum of the values of H for the separated constituent solitons. The result is that the binding potential is *exactly equal to zero* for all the n-solitons. For this reason, they are considered as unstable states. Indeed, an initial perturbation which imparts infinitely small velocities to the constituent solitons will result in splitting. However, this is a slowly growing instability, rather than exponential growth of perturbations, which would imply usual dynamical instability. For this reason, n-solitons may be physically meaningful objects.

In the case of normal GVD, $\beta > 0$, localized (*bright*) solitons do not exist, but a *dark soliton* (DS) is found in this case, in the form of a dark spot ("hole") against a uniform CW background. It is described by the following exact solution to the NLS equation (1.5):

$$u_{DS}(z, \tau) = \frac{\eta}{\sqrt{\gamma}} \tanh\left(\frac{\eta}{\sqrt{\beta}} \tau\right) \exp\left(i\eta^2 z\right), \tag{1.21}$$

where η is an arbitrary amplitude of the background which supports the dark soliton. The DSs are stable, which is possible because the CW background supporting them is itself modulationally stable for $\beta > 0$, as mentioned above. DSs were created experimentally in nonlinear optical fibers [60, 98, 172], about a decade after the first observation of the bright solitons in fibers was reported. DSs are not a subject of this book, except for a brief consideration in the context of the 1D Feshbach-resonance-driven Bose-Einstein condensate, see Fig. 5.3 and related text. A review of DSs can be found in article [91].

An important generalization of the NLS equation is a system of two coupled equations, that describe co-propagation of two waves in an optical fiber. The waves are distinguished by either orthogonal polarizations or different carrier wavelengths. In the general case, the corresponding system is

$$iu_z + icu_\tau - \frac{1}{2}\beta_u u_{\tau\tau} + \gamma \left(|u|^2 + \sigma|v|^2\right) u = 0, \tag{1.22}$$

$$iv_z - icv_\tau - \frac{1}{2}\beta_v v_{\tau\tau} + \gamma \left(|v|^2 + \sigma|u|^2\right) v = 0, \tag{1.23}$$

where σ is the ratio of the SPM and XPM (self-phase-modulation and cross-phase-modulation) coefficients, β_u and β_v are the GVD coefficients (they may be different in the case of two different carrier wavelengths), and a real parameter $2c$ measures the group-velocity mismatch (GVM) between the two waves (in the case of orthogonal polarizations, c accounts for the *group-velocity-birefringence* effect). The cases of different wavelengths or mutually orthogonal circular polarizations correspond to $\sigma = 2$, and two orthogonal linear polarizations are described by Eqs. (1.22) and (1.23) with $\sigma = 2/3$ (strictly speaking, in the latter case the equations also contain *four-wave-mixing* (FWM) nonlinear terms, $(1/3)v^2u^*$ and $(1/3)v^2u^*$, respectively, but they may be usually neglected due to birefringence effects [15]). However, the only case when the system of the coupled NLS equations is integrable (the *Manakov's* system [117]) has $\sigma = 1$.

The system of equations (1.22) and (1.23) conserves the sum of momenta (1.10) in the two components,

$$P_{\text{tot}} \equiv i \int_{-\infty}^{+\infty} uu_\tau^* d\tau + i \int_{-\infty}^{+\infty} vv_\tau^* d\tau. \tag{1.24}$$

The energy (1.9) is conserved separately in each component, unless the FWM terms are included. If the FWM coupling is present, then only the total energy is conserved,

$$E_{\text{tot}} \equiv \frac{1}{2} \int_{-\infty}^{+\infty} |u(\tau)|^2 d\tau + \frac{1}{2} \int_{-\infty}^{+\infty} |v(\tau)|^2 d\tau. \tag{1.25}$$

In the case of $c = 0$, Eqs. (1.22) and (1.23) have obvious two-component (*vectorial*) soliton solutions with $v(z,\tau) = \exp(i\phi_0) u(z,\tau)$ and arbitrary phase shift ϕ_0, that trivially reduce to the ordinary single-component soliton (1.13). In terms of the effective polarization angle θ, such solitons correspond to $\theta = 45°$. The system (1.22), (1.23) with $c = 0$ also gives rise to nontrivial (and stable) soliton solutions with arbitrary polarization ($0 \leq \theta \leq 90°$), which can be found in a numerical form, or in an analytical approximation by means of the VA [87] (only in the Manakov's case, $\sigma = 1$, the vectorial-soliton solutions with $\theta \neq 45°$ can be found in an exact analytical form, with $v(z, \tau) = \exp(i\phi_0)(\tan\theta)u(z, \tau)$).

In the spatial domain, the analysis which leads to solitons is simpler. In this case, relevant solutions to Eq. (1.1) are looked for in the form (1.2), where the amplitude $u(z, x)$ may be a slowly varying function of z and the transverse coordinate x, while the time delay in ϵ is irrelevant, i.e., $\epsilon \equiv \epsilon_0$. The latter implies setting $\omega_0 = k_0/\sqrt{\epsilon_0}$ in the expression for the carrier wave in Eq. (1.2), then the first nontrivial approximation leads to the following nonlinear equation for the slowly varying amplitude,

$$iu_z + \frac{1}{2k_0}u_{xx} + \gamma|u|^2 u = 0, \qquad (1.26)$$

where the relation $I = |u|^2$ is again taken into regard, and this time the nonlinearity coefficient is defined as $\gamma \equiv n_2 k_0/\sqrt{\epsilon_0}$. The spatial-domain equation (1.26) takes exactly the same form (with the same relative signs in front of the second derivative and nonlinear term) as the NLS equation (1.5) in the temporal domain with the anomalous GVD (i.e., the transverse diffraction in the spatial domain is a counterpart to the *negative* GVD in the temporal domain). Accordingly, the family of solutions (1.13), with τ replaced by x, describes spatial solitons, in the form of localized planar beams of light in the two-dimensional plane (z, x). The solutions with $c \neq 0$ correspond to the beams tilted relative to the z axis.

Bragg-grating (*gap*) solitons

The above-mentioned *gap solitons* were experimentally created in a nonlinear optical fiber equipped with a *Bragg grating* (BG) [57], i.e., a periodic system of weak defects in the fiber's cladding, with the period $\lambda/2$, which gives rise to the resonant *Bragg reflection* of the right- and left-traveling electromagnetic waves, with the wavelength λ and local amplitudes $u(x, t)$ and $v(x, t)$, into each other (note that here t is ordinary time, rather than the reduced time τ defined above). A standard model of the BG-equipped nonlinear optical fiber is based on a system of *coupled-mode equations* for the two waves,

$$iu_t + iu_x + \gamma\left(\frac{1}{2}|u|^2 + |v|^2\right)u + \kappa v = 0, \qquad (1.27)$$

$$iv_t - iv_x + \gamma\left(\frac{1}{2}|v|^2 + |u|^2\right)v + \kappa u = 0, \qquad (1.28)$$

where γ is (as above) the nonlinearity coefficient, κ is the Bragg-reflectivity coefficient, and the group velocities of the waves and normalized to be 1. The relative XPM coefficient in Eqs. (1.27) and (1.28) is 2, cf. Eqs. (1.22) and (1.23) for a pair of different wavelengths.

A similar model is known in the spatial domain, with time t replaced by the propagation constant z; in that case, the BG is implemented in the form of a system of parallel grooves (or ridges), with spacing h, on the surface of a planar waveguide, while $u(x, z)$ and $v(x, z)$ are local amplitudes of two waves whose Poynting vectors constitute equal

angles χ with the grooves. The waves are resonantly reflected into each other under
the condition

$$\lambda = 2h \sin \chi, \tag{1.29}$$

where λ is again the wavelength.

Before considering solutions of the full nonlinear equations (1.27) and (1.28), it
is relevant to consider their linearized version (obtained by dropping the cubic terms).
Looking for the corresponding linear-wave solutions as $\{u(x,t), v(x,t) \sim \exp(ipx - i\omega t)\}$,
one immediately finds the corresponding dispersion relation, $\omega^2 = p^2 + \kappa^2$. As seen
from this expression, there are no linear waves whose frequency belongs to the *gap* in
the spectrum (which is also frequently called *bandgap*),

$$-\kappa < \omega < +\kappa. \tag{1.30}$$

Unlike the NLS equation, the system of equations (1.27) and (1.28) is not exactly
integrable (it becomes tantamount to an exactly integrable *massive Thirring model*,
which has been known for a long in the quantum field theory, if the SPM coefficient $\frac{1}{2}$
in the equations is formally replaced by zero). Nevertheless, a family of *exact* soliton
solutions to these equations, with an arbitrary amplitude parameter θ, which takes values $0 < \theta < \pi$ (see below), and an arbitrary velocity c, which is limited to the interval
$-1 < c < +1$, was found in works [13] and [42], following the pattern of the previously known exact solutions in the massive Thirring model. Although the solutions
seem relatively complex, they are quite usable in theoretical analysis:

$$
\begin{aligned}
u_{\mathrm{GS}}(x,t) &= \sqrt{\frac{2\kappa(1+c)}{\gamma(3-c^2)}} (1-c^2)^{1/4} W(X) e^{i\phi(X)-iT\cos\theta}, \\
v_{\mathrm{GS}}(x,t) &= -\sqrt{\frac{2\kappa(1-c)}{\gamma(3-c^2)}} (1-c^2)^{1/4} W^*(X) e^{i\phi(X)-iT\cos\theta}.
\end{aligned} \tag{1.31}
$$

Here, the asterisk stands for the complex conjugation, and

$$
\begin{aligned}
X &= \kappa \frac{x-ct}{\sqrt{1-c^2}}, \quad T = \kappa \frac{t-cx}{\sqrt{1-c^2}}, \\
\phi(X) &= \frac{4c}{3-c^2} \arctan\left[\tanh(X\sin\theta)\tan\left(\frac{\theta}{2}\right)\right], \\
W(X) &= (\sin\theta)\,\mathrm{sech}\left(X\sin\theta - \frac{i\theta}{2}\right).
\end{aligned} \tag{1.32}
$$

The soliton solutions (1.31), (1.32) with zero velocity ($c = 0$), i.e., pulses of *standing
light* pinned by the BG, take an essentially simpler form:

$$
\begin{aligned}
u_{\mathrm{GS}}^{(c=0)}(x,t) &= \sqrt{\frac{2\kappa}{3\gamma}} (\sin\theta) e^{-i(\kappa\cos\theta)t}\,\mathrm{sech}\left(\kappa x\sin\theta - \frac{i\theta}{2}\right), \\
v_{\mathrm{GS}}^{(c=0)}(x,t) &= -\sqrt{\frac{2\kappa}{3\gamma}} (\sin\theta) e^{-i(\kappa\cos\theta)t}\,\mathrm{sech}\left(\kappa x\sin\theta + \frac{i\theta}{2}\right).
\end{aligned} \tag{1.33}
$$

Note that the frequencies of the soliton family (1.33), $\omega_{\text{sol}} = \kappa \cos \theta$, exactly cover the bandgap (1.30), for which reason the solutions themselves are commonly called *gap solitons* (GSs), as mentioned above. This is a manifestation of a very general principle, which states that the linear waves and solitons must occupy different regions in the frequency space. Exceptions to this rule are known too, in the form of *embedded solitons* (they are embedded into the frequency region occupied by the linear waves), which are reviewed in article [39]. However, the embedded solitons, being nongeneric solutions, exist not in families, but at isolated values of the frequency; they also feature very specific stability properties, being *semi-stable* (they are stable in the linear approximation, but, generally speaking, nonlinearly unstable).

An essential difference of the GS solutions from their NLS counterparts (1.13) is a nontrivial phase distribution in the soliton: even in the case of $c = 0$, the solution (1.33) is an essentially complex one, with the intrinsic phases $\pm \arctan \left(\tan \left(\theta / 2 \right) \tanh \left(x \sin \theta \right) \right)$ of its u- and v-components. Another noteworthy difference from the NLS solitons is that the moving GSs cannot be automatically generated from ones with $c = 0$, as Eqs. (1.27) and (1.28) do not obey the Galilean or Lorentz invariance. The GSs are asymptotically equivalent to the NLS solitons only in the limit of $\theta \to 0$, which corresponds to very broad small-amplitude solitons.

Besides the Hamiltonian, Eqs. (1.27) and (1.28) conserve the total momentum and energy, which are given by the same expressions (1.24) and (1.25) as for coupled NLS equations. For the exact GS solutions (1.31), (1.32) they are

$$E_{\text{GS}} \equiv \frac{8\theta \left(1 - c^2 \right)}{\gamma \left(3 - c^2 \right)}, \tag{1.34}$$

$$P_{\text{GS}} = \frac{8\kappa}{\gamma} c \sqrt{1 - c^2} \left[\frac{\left(7 - c^2 \right)}{\left(3 - c^2 \right)^2} \left(\sin \theta - \theta \cos \theta \right) + \frac{\theta \cos \theta}{3 - c^2} \right]. \tag{1.35}$$

The stability of the GSs is quite a nontrivial problem. For the first time, it was considered by means of the variational approximation (VA) in work [113], and later a numerically exact result was obtained in works [28] and [47] by means of numerical computation of the stability eigenvalues, using equations (1.27) and (1.28) linearized for small perturbations. Quite remarkably, the VA had predicted virtually the same result which was later found by dint of the numerical methods: the quiescent solitons ($c = 0$) are stable in slightly more than a half of their existence region, *viz.*, in the interval $0 < \theta < \theta_{\text{cr}} \approx 1.01(\pi/2)$, being unstable against oscillatory perturbations (ones with a complex instability growth rate) in the remaining interval, $1.01(\pi/2) < \theta < \pi$. The stability border, θ_{cr}, very weakly depends on the soliton's velocity c, remaining close to $\pi/2$ up to the limit values of $c = \pm 1$.

Experimental creation of temporal GSs was reported in 1996 [57]. The experiment was run in a short (6 cm) fiber grating. Such a piece of the fiber was sufficient for the formation and observation of the soliton, as the grating induces a very strong artificial dispersion (up to six orders of magnitude stronger than the fiber's natural GVD). To match the strong dispersion, a laser pulse with a very high power was launched into the fiber. The created soliton was observed to travel at the velocity $c \approx 0.75$, in the present notation. In later experiments, the velocity was reduced to ≈ 0.5, while a standing-light soliton, with $c = 0$, has not yet been created in the experiment.

Second-harmonic-generation solitons

Besides the Kerr (alias cubic, or $\chi^{(3)}$) nonlinearity, optical solitons can also be supported by the balance between diffraction/dispersion in the spatial/temporal domain and quadratic ($\chi^{(2)}$), alias second-harmonic-generating (SHG), nonlinearity. Unlike the universal Kerr effect, the $\chi^{(2)}$ nonlinearity occurs only under special conditions in anisotropic media, such as certain optical crystals, or periodically poled waveguides. The existence of $\chi^{(2)}$ solitons was predicted by Karamzin and Sukhorukov back in 1974 [86], but, in the experiment, solitons of this type were created more than 20 years later, first as (2+1)-dimensional spatial solitons (self-supported localized cylindrical beams in a bulk crystal sample) [164], and soon thereafter as spatial (1+1)-dimensional solitons, i.e., localized beams in planar waveguides [155] (the latter are spatial solitons of essentially the same type as described above for the case of the Kerr nonlinearity).

The standard model of the spatial $\chi^{(2)}$ solitons in a planar waveguide includes normalized equations for the local amplitudes $u(x, z)$ and $v(x, z)$ of the fundamental-frequency (FF) and second-harmonic (SH) waves,

$$iu_z + \frac{1}{2}u_{xx} + u^*v = 0,$$

$$2iv_z + \frac{1}{2}v_{xx} + \frac{1}{2}u^2 + qv = 0, \tag{1.36}$$

where x and z have the same meaning as in the spatial-domain model (1.26), i.e., the transverse coordinate and propagation distance, respectively, the nonlinear $\chi^{(2)}$ coefficient is set to be 1, and a real parameter q measures the mismatch between the FF and SH waves. By means of an obvious rescaling, one can always set $q = \pm 1$, for positive and negative q, respectively (or keep $q = 0$, in the case of exact matching).

A single particular solution for the SHG soliton is available in an analytical form for $q = +1$, as shown in the pioneer work by Karamzin and Sukhorukov [86],

$$u(x, z) = \pm \frac{1}{\sqrt{2}}e^{iz/3}\mathrm{sech}^2\left(\frac{x}{\sqrt{6}}\right), \quad u(x, z) = \frac{q}{2}e^{2iz/3}\mathrm{sech}^2\left(\frac{x}{\sqrt{6}}\right). \tag{1.37}$$

A general family of soliton solutions can be sought for in the form of $u(x, z) = e^{ikz}U(x)$, $v(x, z) = e^{2ikz}V(x)$, with $k > 0$ in the case of $q = -1$ or $q = 0$, and $k > 1/4$ in the case of $q = +1$. Except for the single exact solution (1.37) corresponding to $k = 1/3$, the localized functions $U(x)$ and $V(x)$ can be found in a numerical form, or in an approximate analytical form by means of VA, as described in detail in reviews [62] and [35]. The exact solution (1.37) is stable, as well as a larger part of the general soliton family (unstable are only the solitons corresponding to $q = +1$, in a very narrow subregion, $0.25 < k < 0.264$, of their existence region, which is $k > 0.25$).

Experimental observation of $\chi^{(2)}$ solitons in the temporal domain is much more difficult, because of the small propagation distance in available samples, and weak GVD in available materials. Nevertheless, this aim was achieved in an experiment which employed strong artificial dispersion, created by means of a technique using the so-called tilted wave fronts (the technique can be implemented in a waveguide with an extra transverse spatial direction) [49].

A very important property of the quadratic nonlinearity is the fact that it, unlike its $\chi^{(3)}$ counterpart, can support *stable* multidimensional solitons (as mentioned above, the first experimentally created $\chi^{(2)}$ solitons were, effectively, two-dimensional (2D) ones [164]). The problem with the cubic nonlinearity is that it gives rise to *collapse*, i.e., formation of a true singularity after finite evolution, in both the 2D and 3D (three-dimensional) versions of the NLS equation (a detailed account of the collapse theory for the NLS equation can be found in article [29] and book [159]). The possibility of the collapse makes formally existing solitons in both the 2D and 3D equations with the cubic nonlinearity unstable. On the contrary to that, the $\chi^{(2)}$ nonlinearity does not give rise to collapse in any physically relevant dimension, which opens a way to the existence of stable multidimensional solitons. An especially challenging possibility is experimental creation (and possibly use in future applications) of *spatiotemporal solitons* (STSs), alias "light bullets" (the latter term was coined by Silberberg [156]), which are pulses of electromagnetic waves localized in all the directions, transverse and longitudinal. The self-localization in the longitudinal direction actually implies that the soliton is (simultaneously with being a spatial soliton in the transverse directions) a temporal one, as explained above for the solitons in fibers, hence the term STS.

A mathematical model that can generate 3D STSs in SHG media is based on a straightforward generalization of equations (1.36),

$$iu_z - \frac{1}{2}\beta_1 u_{\tau\tau} + \frac{1}{2}\nabla_\perp^2 u + u^* v = 0,$$

$$2iv_z - \frac{1}{2}\beta_2 v_{\tau\tau} + \frac{1}{2}\nabla_\perp^2 v + \frac{1}{2}u^2 + qv = 0, \qquad (1.38)$$

where β_1 and β_2 are the GVD coefficients at the FF and SH, respectively (cf. Eq. (1.5)), and the diffraction operator $\nabla_\perp^2 \equiv \partial^2/\partial x^2 + \partial^2/\partial y^2$ acts on the transverse coordinates. The existence of stable STS solutions to Eq. (1.38) (actually, with $q = 0$) was first predicted, in a rigorous but abstract form (on the basis of variational estimates, without developing any actual approximation to the shape of the soliton) as early as in 1981 by Kanashov and Rubenchik [83]. Similarly, a fully localized 2D STS can exist in a planar waveguide (it is described by Eqs. (1.38) with $\nabla_\perp^2$ replaced by $\partial^2/\partial x^2$).

The first actual approximation for the 3D and 2D spatiotemporal solitons in the generic SHG model, in both analytical and numerical forms (the former was based on the VA), was developed in 1997 [107] (in an earlier paper [97], another approximation was proposed, based on a factorized-product ansatz of the type $u(z, \tau, x) = e^{ikz}F(\tau)G(x)$). This theoretical work was followed by attempts to create STSs in SHG crystals. The best result was the making of effectively 2D "light bullets" of this type in bulk (3D) crystals [103, 101] (the light pulses were self-localized in one transverse direction and in the longitudinal one, while in the other transverse direction they extended across the entire sample). In fact, a full 3D "bullet" could not be created in these experiments, as they employed the above-mentioned tilted-wave-front technique (as the self-localization of the pulse in the temporal direction required sufficiently strong dispersion, that could be induced only artificially), which absorbed one transverse direction. The creation of completely localized STS in three dimensions remains a great challenge to the experiment, and is an intensive for the development of new relevant schemes – first of all, in the theoretical form, before they can be tested in the experiment. An up-to-date review of the topic of optical STSs, including experimental and theoretical aspects, can be found in article [109].

1.1.2 Solitons in Bose-Einstein condensates, and their counterparts in optics

As mentioned above, another (non-optical) field in which solitons have been created quite recently is the Bose-Einstein condensation. A BEC is a low-density vapor of boson atoms of alkali metals (^{87}Rb, ^{23}Na, ^{7}Li, and others) which, by means of special techniques, are cooled down to a temperature on the order of a fraction of nano-Kelvin. In such an ultracold state, all the atoms fall into a single ground state, which is the essence of the condensation (a detailed description of the topic can be found in the book by Pethik and Smith [141]). Existence of solitons in BECs is strongly suggested by the fact that the fundamental equation which describes the evolution of the single-atom wave function $u(x, y, z, t)$ in the condensate, viz., the Gross-Pitaevskii equation (GPE), is similar to its NLS counterpart in optics:

$$i\hbar \frac{\partial u}{\partial t} = \left[-\frac{\hbar^2}{2m} \nabla^2 + U(x, y, z) + \frac{4\pi \hbar^2 a}{m} |u|^2 \right] u, \tag{1.39}$$

where m is the atom's mass, $\nabla^2 \equiv \partial^2/\partial x^2 + \partial^2/\partial y^2 + \partial^2/\partial z^2$ is the usual Laplacian (kinetic-energy operator, in this case), $U(x, y, z)$ (which, in addition, may be time-dependent) is an external potential, and a is the s-wave scattering length which characterizes collisions between the atoms. Positive and negative scattering lengths correspond, respectively, to the repulsion (as in ^{87}Rb) and attraction (as in ^{7}Li) between the atoms.

In particular, it is relevant to consider the BEC of a strongly elongated, i.e., nearly-1D, form, in an appropriately devised magnetic or optical trap which holds the condensate. The nearly 1D trap corresponds to the potential $U = (1/2)m\Omega^2 \left(y^2 + z^2 \right)$, the respective transverse *harmonic-oscillator length*, $l_{\text{ho}} = \sqrt{\hbar/(m\Omega)}$, being much smaller than the longitudinal size of the condensate in the x direction. In this case, the full 3D GPE (1.39) can be effectively reduced to a 1D equation, which is essentially tantamount to the 1D NLS equation with the cubic nonlinearity. The equation looks like Eq. (1.5) with $\beta < 0$, while the sign of the nonlinearity coefficient γ is opposite to that of the scattering length a. Thus, effectively one-dimensional bright *matter-wave solitons*, similar to the temporal solitons found in nonlinear optical fibers, may be naturally expected in the elongated condensate if the atomic interactions are attractive, with $a < 0$. Indeed, bright solitons in the ^{7}Li condensate, in which the atoms interact attractively (but very weakly, which makes it possible to prevent collapse of the condensate, that would be inevitable in the 2D or 3D case), were created in two independent experiments [158, 92]. Still before that, dark solitons were created in repulsive condensates (rubidium and sodium) [34, 46].

The most recent achievement in this direction is the experimental creation of weakly localized 1D bright solitons in an (effectively one-dimensional) *repulsive* condensate of ^{87}Rb loaded into a periodic potential in the form of an *optical lattice* (OL; it is an ordinary interference pattern created by two coherent laser beams that illuminate the condensate from opposite directions [58]). The OL corresponds to the longitudinal potential $U(x) = \epsilon \cos(kx)$ in the GPE (1.39), as well as in its reduced 1D counterpart. Despite the fact that the repulsive cubic nonlinearity cannot create bright solitons in the free space, its interplay with the periodic OL potential can support bright solitons of the *gap type*, as first predicted by Baizakov, Konotop and Salerno [23] (see also papers [135] and [56]). To understand this possibility, one should note that, as is well known, the 1D linear Schrödinger equation with the periodic potential, $\epsilon \cos(kx)$, gives rise to a spectrum with finite bandgaps separating distinct Bloch bands that host linear-wave spatially quasi-periodic solutions. The repulsive cubic nonlinearity may support gap solitons (GSs) with frequencies belonging to these finite bandgaps. To understand this possibility in simple terms, one may notice that the fiber-BG model (1.27), (1.28) supports the family of the gap-soliton solutions (1.31) – (1.33) irrespective of the overall sign in front of the cubic terms, as the sign reversal may be compensated by complex conjugation of the equations. The GSs of this type extend over many cells of the lattice potential (they are weakly localized in that sense) and feature the wave function $u(x)$ that oscillates (in x), many times crossing zero and gradually vanishing at $|x| \to \infty$.

In the multidimensional case, the attractive cubic nonlinearity in the GPE (with $a < 0$), as well as in its NLS counterpart in optics, gives rise to collapse. Nevertheless, the 2D or 3D periodic potential of the OL type in Eq. (1.39), i.e.,

$$U(x, y, z) = \epsilon \left[\cos(kx) + \cos(ky) + \cos(kz) \right] \tag{1.40}$$

(in the 3D case), can readily stabilize the corresponding multidimensional solitons against collapse [24]. A solution for the soliton is looked for as

$$u(x, y, z, t) = e^{-i\mu t/\hbar} v(x, y, z), \tag{1.41}$$

where the constant μ is a real chemical potential (in similar optical models, it would be the propagation constant), and the real function $v(x, y, z)$ satisfies the stationary equation,

$$\mu v = \left[-\frac{\hbar^2}{2m} \nabla^2 + U(x, y, z) + \frac{4\pi\hbar^2 a}{m} v^2 \right] v, \tag{1.42}$$

Depending on the value of the soliton's norm (which measures the number of atoms in the condensate),

$$N_{2D} = \int\int [v(x, y)]^2 \, dxdy, \quad N_{3D} = \int\int\int [v(x, y, z)]^2 \, dxdydz, \tag{1.43}$$

the stable multidimensional solitons may assume a "single-cell" shape, being confined essentially to a single cell of the underlying OL potential (1.40), or a multi-cell form, see typical examples in Fig. 1.1.

In the multi-dimensional GPE with the OL potential and *repulsive* nonlinearity ($a > 0$), weakly localized stable bright solitons of the gap type can be created by essentially the same mechanism which, as mentioned above, gives rise to the GSs in the 1D case. This possibility was for the first time predicted, together with the 1D gap solitons, in the above-mentioned work [23]. An example of the 2D gap soliton is displayed in Fig. 1.2(a).

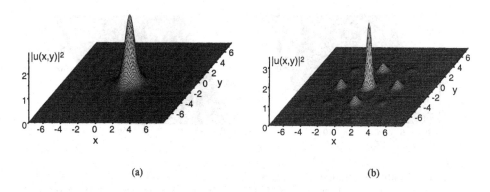

(a)   (b)

Figure 1.1: Typical examples of stable *single-cell* (a) and *multi-cell* (b) solitons in the 2D Gross-Pitaevskii equation with the attractive nonlinearity and optical-lattice potential. The ratio of the strengths of the optical-lattice potential in the cases (a) and (b) is $\epsilon_a/\epsilon_b = 0.092$, and the ratio of the corresponding norms (see Eq. (1.43)) is $N_a/N_b = 1.98$.

In fact, the dimension of the OL which is sufficient for the stabilization of the multidimensional solitons in the GPE with the attractive potential is smaller by 1 than the dimension of the equation itself, as it was independently shown in the works [25, 26] and [125]: a quasi-2D potential, given by the expression (1.40) without the last term, $\cos(kz)$, can support stable fully localized 3D solitons. Similarly, a quasi-1D potential (the one containing only the term $\cos(kx)$) can stabilize a fully localized 2D soliton, but it cannot stabilize a 3D soliton [25, 26]. Naturally, the shape of the soliton in such a *low-dimensional lattice* is strongly anisotropic, in the directions across and along the lattice, as illustrated by a typical example of the 3D stable soliton displayed in Fig. 1.3.

Besides the fundamental 2D solitons outlined above (ones with the peak at the center), the GPE stabilized by the OL potential can also give rise to 2D solitons with *embedded vorticity S*, where S is an integer, $S = 0$ corresponding to the fundamental solitons. The presence of the vorticity means that the phase of the stationary complex soliton solution acquires a change of $2\pi S$ along a closed path surrounding the soliton's center. The concept of vortices is a very well-known one (see a book [145]), but it is usually considered in isotropic media, and in that case the *topological charge* (alias vorticity), S, is a dynamical invariant, whose conservation is tantamount to the conservation of the angular momentum. Obviously, the OL potential breaks the isotropy, hence the vorticity cannot be an integral of motion of the GPE (1.39). Nevertheless, the vorticity can be defined for a given stationary solution, even in an anisotropic model. As it was independently shown in works [24] and [174], the vortices with $S = 1$ are stable in the 2D GPE (1.39) with the attractive nonlinearity ($a < 0$); vortices with higher values of S may be stable too. In the equation of the same type with the repulsive cubic term ($a > 0$), stable 2D solitons of the gap type can also carry embedded vorticity, as

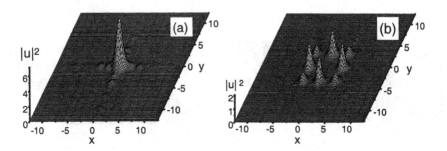

Figure 1.2: Stable 2D fundamental gap soliton (a) and its vortical counterpart with the topological charge $S = 1$ (b), in the 2D Gross-Pitaevskii equation with the repulsive nonlinearity and optical lattice. Both solitons have the same norm N_{2D}, and are found for the same strength and period of the optical-lattice potential.

was shown independently in several works [25, 152, 136]. A typical example of such a gap-soliton vortex is displayed in Fig. 1.2(b).

The GPE in two dimensions, supported by periodic potentials, has its important counterpart in nonlinear optics, in the form of models describing *photonic crystals* and *photonic-crystal fibers* (PCFs, alias *microstructured fibers*). Fabrication of the first PCF was reported in 1996 [94]. Generally, it may be realized as a "thick fiber" with an intrinsic structure in the form of a system of voids running parallel to the fiber's axis. In the transverse plane, the voids form a regular lattice (most frequently, a hexagonal one), which frequently includes a relatively wide empty core in the center. The respective NLS equation, governing the spatial evolution of the local amplitude of the electromagnetic field in the coordinate z running along the PCF's axis, is similar to the GPE (1.39) with the coordinates x and y, where t is replaced by z. In this case, a 2D periodic potential, resembling the 2D version of the expression (1.40), accounts for the periodic modulation of the refractive index in the PCF's transverse plane due to the fiber's microstructure. The difference from Eq. (1.39) is that the nonlinearity coefficient in a PCF model is also subject to a periodic modulation in x and y, as the inner holes have no nonlinearity. Similar to the 2D GPE with the OL potential, the NLS equation for the PCF supports stable spatial solitons (localized in x and y and uniform along z) [173, 64] and soliton vortices [65], that were found in direct numerical simulations.

Another counterpart of the 2D GPE with the periodic OL potential describes spatial solitons in a photorefractive optical medium, in which an effective photonic lattice, in the coordinates x and y, is induced by an interference pattern produced by coherent laser beams, with large intensity I_0, illuminating the crystal in the ordinary polarization, in which the medium is nearly linear. Then, a signal beam is launched in the extraordinary polarization, which feels a mixture of a strong *saturable* nonlinearity and the *virtual photonic lattice* induced by the transverse illumination. The corresponding dynamical model is based on the following equation for the local amplitude $u(x, y, z)$

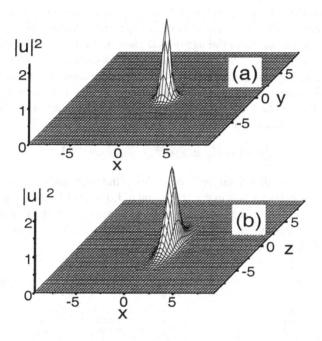

Figure 1.3: An example of a stable single-peaked 3D soliton in the Gross-Pitaevskii equation (1.39) with the attractive nonlinearity ($a < 0$) and quasi-2D optical-lattice potential (1.40) (i.e., one without the term $\cos(kz)$). The 3D soliton is shown through its 2D cross sections, in the planes $z = 0$ (a) and $y = 0$ (b). Recall that the quasi-2D optical-lattice potential does not depend on z, but depends on x and y.

of the signal beam, confined in the (x, y) plane and evolving along z,

$$iu_z + \frac{1}{2}\nabla^2 u - \frac{u}{1 + I_0[\cos(kx) + \cos(ky)]^2 + |u|^2}, \qquad (1.44)$$

cf. Eqs. (1.39) and (1.40).

As well as the above-mentioned PCF model, the one described by Eq. (1.44) gives rise to stable 2D solitons, as was first predicted in work [55]. These objects have already been created in the experiment – first, fundamental solitons [66], and then vortical ones [132, 67].

The 1D version of Eq. (1.44), with the expression $[\cos(kx) + \cos(ky)]^2$ in the denominator replaced by $\cos^2(kx)$, is a physically relevant model too, applying to the propagation of the signal wave through the 1D photonic lattice. In this case, assuming a strong lattice, i.e., $I_0 \gg 1$, the full equation with the x-dependent coefficients may be approximated by a system of two coupled equations with constant coefficients, within the framework of the *coupled-mode approximation* – similar to that which leads to the derivation of the standard GS model (1.27), (1.28). To this end, one introduces the couple-mode representation of the field,

$$u(x, z) = U_+(x, z)e^{ikx} + U_-(x, z)e^{-ikx}, \qquad (1.45)$$

where U_+ and U_- are slowly varying amplitudes of the right- and left-traveling waves. Substituting this representation in the 1D variant of Eq. (1.44), Fourier expanding over the harmonic set of $\exp(inkx)$ with integer n, and keeping only the fundamental harmonics ($n = \pm1$) leads to the *couple-mode system* with a nonlinear saturable coupling [108], cf. Eqs. (1.27) and (1.28).

$$
\begin{aligned}
i\frac{\partial U_+}{\partial z} + ik\frac{\partial U_+}{\partial x} &= \frac{U_+ - U_-}{\sqrt{I_0(1 + |U_+ - U_-|^2) + 1 + 2(|U_+|^2 + |U_-|^2)}}, \\
i\frac{\partial U_-}{\partial z} - ik\frac{\partial U_-}{\partial x} &= \frac{U_- - U_+}{\sqrt{I_0(1 + |U_+ - U_-|^2) + 1 + 2(|U_+|^2 + |U_-|^2)}}.
\end{aligned} \qquad (1.46)
$$

In fact, one combination of these equations is just a linear relation,

$$\frac{\partial}{\partial z}(U_+ + U_-) + k\frac{\partial}{\partial x}(U_+ - U_-) = 0. \qquad (1.47)$$

Equations (1.46) are much more convenient, than the underlying equation (1.44), for the analysis of soliton solutions [108].

1.2   The subject of the book: solitons in periodic heterogeneous media ("soliton management")

1.2.1   General description

With all the profound importance of the basic exactly integrable systems, such as the KdV, modified KdV, NLS, derivative-NLS, sine-Gordon equations and others, they are exceptional models, in the sense that any additional term, which takes into regard physical effects that were not included in the basic model, breaks the exact integrability. This circumstance suggest the necessity to investigate solitons in nonintegrable models (in a strict mathematical sense, these solutions are not "solitons", but rather "solitary waves"; however, following commonly adopted practice, they will be called solitons). In many physically relevant situations, the additional terms which break the integrability of the unperturbed model are small, making it natural to apply a perturbation theory relying on an asymptotic expansion around exact solutions provided by the IST in the absence of the perturbations. Numerous results generated by this technique were reviewed in article [93].

There is another technique, based on the *variational approximation* (VA), which is less accurate than the one based on the IST, but applicable to a much broader class of models, as it only requires a possibility to derive the corresponding equation from a Lagrangian (i.e., the equation must have a variational representation), and does not rely upon proximity to any integrable limit. For instance, in the case of the GPE, the VA approximates the mean-field wave function, as a function of spatial coordinates, by a particular analytical expression (*ansatz*) which contains several free parameters that may be functions of time. Numerous examples of *ansätze* (plural for ansatz) are considered below, see Eqs. (2.6), (1.51), (2.31), (5.16), (7.5), (8.5), (9.6), (9.15), (9.18), (10.3). Effective evolution equations for the free parameters are derived using the Lagrangian representation of the underlying PDE. For the NLS solitons in one dimension, this technique was first proposed by Anderson [19], and it was applied for the first time to BEC in works [142, 143]. Many results generated by VA were collected in an extensive review article [104].

Another approximate technique, which also applies to a broad class of models, is based on the method of moments. In the general case, this approach too relies upon approximating the wave function (in the case of BEC) by an ansatz with a fixed functional dependence on the spatial coordinates, and several free parameters that may vary in time (see, e.g., papers [71] and [6] for the application of this method to the GPE, and paper [144] for an application to nonlinear fiber optics). A system of evolution equations for free parameters is derived by multiplication of the PDE by several appropriately chosen functions of the spatial coordinates (the number of functions must be equal to the number of the free parameters). Each time, the result of the multiplication is explicitly (analytically) integrated over the spatial coordinates, which casts it in the form of an ODE. In the case of the two-dimensional GPE with the isotropic parabolic potential, an *exact closed system* of evolution equations for moments can be derived [71].

A vast class of nonintegrable but physically important models includes a periodic modulation of some of the system's parameters in space and/or in time (early results obtained by means of the perturbation theory for solitons in weakly inhomogeneous media were reviewed in article [93]; still earlier, some results were collected in book [8]). In particular, it was explained above that a spatially periodic potential, such as one generated by the OL in BEC, see Eqs. (1.39) and (1.40), or by the photonic lattice in the photorefractive crystal, see Eq. (1.44), can easily stabilize multidimensional solitons, which is an example of a principally important effect produced by the *transverse modulation*, that does not involve the evolutional variable. Another possibility is *longitudinal modulation*, which makes coefficients of the governing equation periodic functions of the evolutional variable, i.e., the propagation distance z, in optical models of the NLS type (see Eqs. (1.5), (1.26), and (1.44)), or time t, in the GPE (1.39). A drastic difference of the longitudinal modulation from its transverse counterpart is that the corresponding mathematical problem is a *non-autonomous* one. As concerns solitons, their existence in transversely modulated models is quite obvious, as the substitution separating the evolutional and transverse variables, such as in Eq. (1.41), leads to an equation of the same type as Eq. (1.42), which should have soliton solutions, in the general case. On the contrary, in models subjected to the longitudinal modulation the very existence of robust soliton solutions is a highly nontrivial issue, as is explained below and in the rest of the book.

1.2.2   One-dimensional optical solitons

A paradigmatic system featuring the longitudinal modulation is a long fiber-optic telecommunication link subjected to the *dispersion management* (DM), i.e., periodic alternation of positive and negative values of the GVD coefficient along the link. This means that it is built as a periodic concatenation of two different species of optical fibers, with opposite values of the GVD coefficient (as described in review [168]). Accordingly, the corresponding NLS equation differs from its standard form (1.5) in that the GVD coefficient is a periodic function of z:

$$iu_z - \frac{1}{2}\beta(z)u_{\tau\tau} + \gamma|u|^2 u = 0. \tag{1.48}$$

In the practically important case, this coefficient is a piecewise-constant function,

$$\beta = \begin{cases} \beta_1 > 0, & \text{if } 0 < z < L_1, \\ \beta_2 < 0, & \text{if } L_1 < z < L_1 + L_2 \equiv L, \end{cases} \tag{1.49}$$

which repeats with a period L. This form of the periodic modulation of the GVD coefficient is usually referred to as the *DM map*. Note that, in a more realistic model, the coefficient γ in Eq. (1.48) also periodically jumps between two different values, corresponding to the two alternating species of the fiber, but this feature is less significant, as the nonlinearity coefficient does not change it sign.

Transmission of solitons, or, more generally, pulses localized in the temporal variable τ (in optical telecommunications, they are commonly referred to as *return-to-zero* (RZ) pulses, which implies that the field $u(z, \tau)$ becomes very close to zero between the pulses), through long DM systems is an issue of fundamental importance to applications. The reason is that all the existing medium- and long-haul commercial telecommunication fiber-optic networks are built in the *dispersion-compensated* form, which corresponds to Eq. (1.49), with the *path-average dispersion* (PAD),

$$\beta_0 = \frac{\beta_1 L_1 + \beta_2 L_2}{L_1 + L_2}, \tag{1.50}$$

equal to zero or very small. This is necessary because, in the usual linear regime, in which the networks actually operate (with the exception of the single soliton-based commercial link, about 3,000 km long, which was built in Australia in 2002 [126]), RZ signals avoid systematic degradation only if $\beta_0 = 0$. Indeed, the linear version of Eq. (1.48) (with $\gamma = 0$) gives rise to an exact RZ solution in the form of a Gaussian pulse,

$$u_{\text{RZ}}^{(\text{linear})}(z, \tau) = \sqrt{\frac{P_0}{1 - 2iB(z)/W_0^2}} \exp\left(-\frac{\tau^2}{W^2(z)} - \frac{2iB(z)}{W_0^2 W^2(z)}\right), \quad (1.51)$$

where the *accumulated dispersion* is defined as

$$B(z) \equiv B_0 + \int_0^z \beta(z')dz' \quad (1.52)$$

(B_0 is an arbitrary constant determined by initial conditions), the temporal width of the RZ pulse is

$$W(z) = \sqrt{W_0^2 + \frac{4B^2(z)}{W_0^2}}, \quad (1.53)$$

while W_0 and P_0 are arbitrary parameters that measure the width and peak power of the pulse (the width is measured at points where $B(z) = 0$, which means that the pulse is narrowest at these points). As seen from Eq. (1.53), the width of the pulse evolves in the course of the propagation, and the pulse suffers no systematic degradation, i.e., the width does not grow on average, solely in the case when the mean value of $\beta(z)$, i.e., PAD (1.50), vanishes, $\beta_0 = 0$. Thus, if the operation regime is to be upgraded by replacing linear signals by nonlinear RZ pulses, it is very important to consider the transmission of nonlinear pulses in the dispersion-compensated links.

Getting back to the full NLS equation (1.48) with the variable GVD coefficient, it is relevant to mention that it admits the same Hamiltonian representation (1.7) and (1.8) as above, but the Hamiltonian with the z-dependent coefficient $\beta(z)$ is no longer a dynamical invariant. Nevertheless, the energy (1.9) and momentum (1.10) remain dynamical invariants in this case.

The DM is just the first example of nonlinear *periodic heterogeneous systems* which feature the longitudinal modulation. Generally speaking, one may expect that a soliton, periodically passing from one segment of the system into another, with strongly differing parameters, will be quickly destroyed. In the most general case, this is true indeed. Nevertheless, a nontrivial fact is that there is a *class of systems* where very robust solitons can be found, despite the fact that they propagate through a strongly heterogeneous structure. In an explicit form, a concept of such a class of the soliton-bearing periodic systems was for the first time formulated in paper [85], where another realization of this class was reported, in the form of a model supporting robust spatial solitons in a channel which is heterogeneous in the longitudinal direction. The channel is built as a periodic concatenation of waveguiding and anti-waveguiding segments, which share the self-focusing Kerr ($\chi^{(3)}$) nonlinearity.

Further, the concept of the DM suggests its counterpart in the form of *nonlinearity management* (NLM), i.e., insertion (into a long fiber-optic telecommunication link) of special nonlinear elements that can compensate the accumulated nonlinear phase shift generated by the Kerr effect in the fiber. This possibility was first proposed, in a rather abstract form, in work [139], and was further developed in papers [78, 54], where it was demonstrated that SHG modules can be used to generate an effective *negative* Kerr effect (through the so-called *cascading mechanism*, i.e., repeated action of the corresponding quadratic nonlinearity), which will play the compensating role. The NLM model is described by equation (1.48), in which both $\beta(z)$ and $\gamma(z)$ periodically jump between positive and negative values.

Still another example of periodic heterogeneous systems is provided by the *split-step model* (SSM), that was introduced in work [50] as a periodic concatenation of nonlinear dispersionless and linear dispersive segments. The term SSM stems from the name of the well-known numerical technique used for simulations of the NLS equation and similar equations, which splits each step of marching forward into two sub-steps, one purely nonlinear, and the other one purely dispersive. Unlike the numerical scheme, in the SSM proper this separation is not an artificial trick, but a real physical feature, as the lengths of the segments are not small, but are rather comparable to the characteristic nonlinearity- and dispersion lengths of the pulses, such as the one given by Eq. (1.16). Quite surprisingly, the RZ pulses, which may also be called solitons in this case, were found to be very robust in the SSM, in a fairly large region in the corresponding parameter space. They are robust too if a small amount of nonlinearity is added to dispersive segments, and weak dispersion is admixed to the nonlinear ones (in fact, such a system is a "hybrid" of the SSM and DM) [53], as well as in a multi-channel generalization of the SSM [51]. The latter system implements the WDM (wavelength-division-multiplexed) scheme, which is the basis of the operation of fiber-optic telecommunications networks. Moreover, it was also found that both DM and SSM solitons remain stable in the presence of loss (natural absorption in the fiber) and compensating amplification.

Another model, which combines the NLM with the effective dispersion (actually, diffraction, as the model was realized in the spatial domain) induced by the Bragg grating, was investigated too [21]. It is composed of alternating BGs with opposite signs of the Kerr nonlinearity, and also gives rise to a family of robust solitons.

A periodic heterogeneous system with the $\chi^{(2)}$ nonlinearity was proposed in the form of the so-called *tandem model* [162], which combines linear segments and ones carrying the quadratic nonlinearity [162]. Specific solitons were revealed by numerical simulations in this model.

In all these systems, stable solitons exist in the form of periodically oscillating *breathers* (obviously, the solitons cannot keep a permanent shape propagating through the inhomogeneous structure). Qualitatively, the breathers resemble the exact solution (1.51) in the linear DM model, found under the above-mentioned condition $\beta_0 = 0$. The periodic oscillations make theoretical analysis of the solitons and their stability essentially more complex than in the case of the ordinary stationary solitons.

To complete the general discussion of the periodically managed 1D optical solitons, it is relevant to mention that quite robust solitons can also be found in *randomly*, rather than periodically, modulated systems. A practically important example is provided by stable solitons found in the model of random DM, which is described by the above equations (1.48) and (1.49), with a difference that the length L of the DM map varies randomly from a cell to a cell [106]. This situation corresponds to real terrestrial telecommunication networks. Another example is a *random SSM*, in which, too, the length of the system's cell varies randomly along the propagation distance [52].

1.2.3 Multidimensional optical solitons

All the above examples of nonlinear periodic heterogeneous systems supporting stable solitons are one-dimensional. Multidimensional systems which may be included into the same general class were found too. An essential example is a model of a bulk medium composed of alternating layers with self-focusing and self-defocusing Kerr nonlinearity [165], which may be regarded as a multidimensional counterpart of NLM systems. Stable 2D cylindrical solitons were found in this model, both in the numerical form and by means of VA, which implies that the periodic alternation of the self-focusing and self-defocusing suppresses the instability against collapse. However, 3D "light bullets" (STSs) are unstable in this setting, as well as 2D vortical solitons [165].

The DM model also has its 2D counterpart, constructed by adding a transverse coordinate to the temporal-domain equation (1.48). Stable solitons, both single- and doubled-peaked ones, were found in this 2D model, but, again, the DM alone cannot stabilize 3D solitons ("bullets") [122].

The above-mentioned 1D model of the tandem type, with alternating linear and $\chi^{(2)}$ segments, can also be made two-dimensional, and stable light-bullet solutions exist in it [163]. In this case, the stabilization of the 2D soliton is more straightforward, as the $\chi^{(2)}$ nonlinearity does not give rise to collapse.

1.2.4 Solitons in Bose-Einstein condensates

Mechanisms admitting "management" of solitons were also developed in BEC models. In particular, the size and *sign* of the scattering length a in GPE (1.39) can be, in some cases, easily controlled by external magnetic field, through the effect of the *Feshbach resonance* (FR), as was predicted theoretically [82] (FR can be also induced and controlled by an external optical field [61]), and then demonstrated in direct experiments in BECs [81, 148, 157].

The external magnetic field which gives rise to the FR can be made time-periodic, with the zero mean value, which induces periodic harmonic modulation of the nonlinearity coefficient in the GPE. In the 1D situation, this opens a way to develop a technique of the *Feshbach-resonance management* (FRM), as was proposed in work [90] (below, this mode of handling the condensate will be sometimes called "ac-FRM" control, to stress that the nonlinearity coefficient in the corresponding GPE periodically changes its sign). Actually, FRM is a BEC counterpart of the above-mentioned nonlinearity management in fiber-optic links, with the propagation distance z replaced (as the evolutional variable) by time t. It is still more interesting that the FRM can stabilize solitons against collapse in the GPE in two dimensions [5, 150], which is a direct counterpart of the above-mentioned stabilization mechanism for the 2D spatial optical solitons in a layered material with periodic alternation of the sign of the Kerr coefficient. It is noteworthy too that, as well as in the optical model, 3D solitons cannot be made stable by means of the FRM technique alone (a recent result [166] is that the stabilization of solitons in the 3D space is possible if the FRM is combined with a one-dimensional OL – recall that a 1D lattice alone cannot stabilize 3D solitons either).

1.2.5   The objective of the book

As was outlined above, certain understanding has been accumulated in the study of one- and multi-dimensional solitons which turn out to be stable as they propagate through a nonlinear medium periodically modulated in space in the longitudinal direction, or evolve under the action of a time-periodic field. Moreover, stable solitons of approximately the same type can sometimes be found even in the case when the spatial or temporal "management" of the system is random, rather than periodic.

Unlike integrable models, no rigorous or truly general method (such as the IST or bilinear Hirota representation) is known for analysis of the soliton dynamics in this class of the longitudinally modulated systems. Nevertheless, it is possible to collect essential results obtained by means of numerical and, in some cases, (semi-) analytical methods in particular models falling into the class of the periodic heterogeneous nonlinear systems (including some random systems), and arrive at sufficiently general conclusions concerning fundamental properties of solitons in such systems. This is the objective of the book.

Chapter 2

Periodically modulated dispersion, and dispersion management: basic results for solitons

2.1 Introduction to the topic

Dispersion management (DM) is a name commonly adopted in the literature for the model based on the NLS equation (1.48) with a constant nonlinearity coefficient γ and the sign-changing GVD coefficient β, modulated along the propagation distance z as per Eq. (1.49). As explained in Introduction, the transmission of quasi-soliton signals, alias *return-to-zero* (RZ) pulses, in the DM model is an issue of fundamental importance to fiber-optic telecommunications. It should be stressed that, unlike many other nonlinear systems with periodic management, where the results have thus far been chiefly theoretical, the DM solitons in fiber-optic telecommunication links were studied in the experiment in detail, see, e.g., paper [38].

Most theoretical works which studied the soliton transmission in the DM model relied on direct numerical simulations. As concerns analytical approaches, two most significant ones are based on the variational approximation (VA), and on integral equations. Actually, both methods assume that the nonlinearity in the model is weak enough, so that, in the zero-order approximation, the RZ pulse may be approximated by the expression (1.51), which is an exact solution to the linear version of Eq. (1.48).

The integral formalism for the DM solitons was worked out by Gabitov and Turitsyn [69] and Ablowitz and Biondini [9] (see also paper [138]). It is based on an idea that, in the linear limit, a general solution to equation (1.48) can be searched for by means of the Fourier transform as

$$u(z, \tau) = \frac{1}{2\pi} \int_{-\infty}^{+\infty} e^{-i\omega\tau} \hat{u}(z, \omega) d\omega. \tag{2.1}$$

Substituting this representation in the linear version of Eq. (1.48), one immediately derives an evolution equation for the Fourier transform:

$$\frac{d\hat{u}}{dz} = \frac{i}{2} \omega^2 \beta(z) \hat{u}, \tag{2.2}$$

a solution to which is obvious,

$$\hat{u}(z, \omega) = \hat{u}(0, \omega) \exp\left(\frac{i}{2} \omega^2 B(z)\right), \tag{2.3}$$

where $B(z)$ is the *accumulated dispersion*, defined as per Eq. (1.52).

If the nonlinearity is taken into regard, the wave field can still be represented in the form of Eq. (2.1), but then the nonlinear term in Eq. (1.48), after the substitution of the Fourier representation, will add a cubic integral term to the evolution equation (2.2). Various results can then be obtained from the analysis of the nonlinear integral equation.

Similar to the integral formalism, the VA also makes use of the linear limit of Eq. (1.48); however, it starts not with the general linear superposition (1.51), but rather with the fundamental solution (1.51). The general idea of the VA is that, after the introduction of the weak nonlinearity, the Gaussian wave form (1.51) may be an adequate *ansatz* for the solution, assuming that its constant parameters may become slowly varying functions of z; the main objective of the variational technique is to derive equations governing slow evolution of the parameters. In this chapter, the variational approach will be presented following, chiefly, paper [100].

Besides the model with the DM map (1.49), it is also interesting to consider a system with the harmonic modulation of the GVD coefficient,

$$\beta(z) = -(1 + \varepsilon \sin z), \tag{2.4}$$

where the PAD is normalized to be -1, and the modulation period is scaled to be 2π. Although the sinusoidal modulation is not a realistic assumption for fiber-optic telecommunications, the model is of interest in its own right, as it may predict quite interesting results even for relatively small values of the modulation amplitude ε, when the local GVD coefficient (2.4) does not change its sign, remaining always negative (i.e., corresponding to the anomalous GVD). In fact, nontrivial results (such as splitting of a soliton into two, see below) may be generated by resonances between internal vibrations of a perturbed soliton, and the periodic modulation defined by Eq. (2.4).

In this chapter, the analysis will be performed first for the model (2.4), and then for the one (1.49). The VA (combined with direct simulations) will be used in both cases, but the results will be very different, due to the fundamental difference between the two types of the periodic modulation.

2.2 The model with the harmonic modulation of the local dispersion

In the case of the harmonic modulation (2.4), the NLS equation (1.5) takes the form

$$iu_z + \frac{1}{2}\left(1 + \varepsilon \sin z\right) u_{\tau\tau} + |u|^2 u = 0, \tag{2.5}$$

where the normalization $\gamma = 1$ is adopted. This model was introduced in 1993 in paper [110], with the intention to study possible resonances in it. In that first work, only the VA was used, without direct simulations. An important contribution to the analysis of the model was later made by Abdullaev and Caputo [3], and direct comparison of the predictions of the VA with direct simulations, which reveal effects that the VA could not predict, was reported in paper [76].

2.2.1 Variational equations

The application of the VA to the soliton in the model (2.5) was described many times and summarized in review [104], therefore here it will be presented in a brief form. The VA assumes the ansatz which mimics the exact NLS soliton solution (1.13), but with arbitrary amplitude A, temporal width a, and phase ϕ; in addition, it is assumed that the nonstationary soliton may have *chirp*, i.e., a parabolic phase profile across the pulse, with a real coefficient b in front of it:

$$u_{\text{ansatz}}(z, \tau) = A(z)\operatorname{sech}\left(\frac{\tau}{a(z)}\right)\exp\left[i\phi(z) + ib(z)\tau^2\right]. \tag{2.6}$$

All the free parameters in the ansatz are allowed to be functions of the evolutional variable z, the first objective being to derive a system of evolution equations for them. This is done using the fact that the NLS equation can be derived, in the form of $\delta S/\delta u^* = 0$, from the *action functional* $S\{u, u^*\}$, where $\delta/\delta u^*$ is the symbol for the variational derivative. The action is represented in the form of $S = \int L\,dz$, where L is a *Lagrangian*, that has its own integral form, $L = \int_{-\infty}^{+\infty} \mathcal{L}\,d\tau$, with a *Lagrangian density* $\mathcal{L}$, which must be real. For the NLS equation (1.5), the latter is

$$\mathcal{L} = \frac{i}{2}\left(u^* u_z - u u_z^*\right) + \frac{1}{2}\beta(z)|u_\tau|^2 + \frac{1}{2}\gamma|u|^4. \tag{2.7}$$

The insertion of the ansatz (2.6) into the Lagrangian and analytical integration in τ yields the corresponding effective Lagrangian, as a function of the variational parameters and their z-derivatives (denoted by the prime),

$$L_{\text{eff}}^{\text{(NLS)}} = -2A^2 a\phi' - \frac{\pi^2}{6}A^2 a^3 b' + \frac{1}{3}\beta(z)\frac{A^2}{a} - \frac{\pi^2}{3}DA^2 a^3 b^2 + \frac{2}{3}\gamma A^4 a. \tag{2.8}$$

A standard set of the variational (Euler-Lagrange) equations follows from the effective Lagrangian,

$$\frac{d}{dz}\frac{\partial L_{\text{eff}}^{(\text{NLS})}}{\partial \phi'} = 0, \quad \frac{d}{dz}\frac{\partial L_{\text{eff}}^{(\text{NLS})}}{\partial b'} - \frac{\partial L_{\text{eff}}^{(\text{NLS})}}{\partial b} = 0,$$

$$\frac{\partial L_{\text{eff}}^{(\text{NLS})}}{\partial a} = \frac{\partial L_{\text{eff}}^{(\text{NLS})}}{\partial A} = 0. \tag{2.9}$$

After straightforward manipulations, these equations can be cast in the following form:

$$\frac{d}{dz}\left(A^2 a\right) = 0, \tag{2.10}$$

$$b = -\left(2\beta(z)a\right)^{-1}\frac{da}{dz}, \tag{2.11}$$

$$\frac{d}{dz}\left[(-\beta(z))^{-1}\frac{da}{dz}\right] = -\frac{\partial U_{\text{eff}}(a)}{\partial a}, \tag{2.12}$$

$$U_{\text{eff}}(a) \equiv -\frac{2}{\pi^2}\left(\beta a^{-2} + 2\gamma E a^{-1}\right), \quad E \equiv A^2 a, \tag{2.13}$$

supplemented by a separate equation for the phase,

$$\frac{d\phi}{dz} = \frac{\pi^2}{12}a^2\left[\frac{db}{dz} - 2\beta(z)b^2\right] - \frac{1}{6}\frac{\beta(z)}{a^2} - \frac{2}{3}\gamma A^2. \tag{2.14}$$

Equation (2.10) implies the existence of the dynamical invariant $E \equiv A^2 a$. The conservation of this quantity is a straightforward manifestation of the conservation of energy (1.9) in the full NLS equation. Indeed, the substitution of the ansatz (2.6) into Eq. (1.9) yields $E = A^2 a$. Equation (2.11) shows that the intrinsic chirp of the soliton is generated by its deformation (change of the width).

Equations (2.12) and (2.13) demonstrate that the evolution of the soliton's width can be represented as a motion of a Newtonian particle with a variable mass $-1/\beta(z)$ and a coordinate $a(z)$ in the potential well $U_{\text{eff}}(a)$, while the propagation distance z plays the role of time. For the case when β is a negative constant, the potential is shown in Fig. 2.1. In this case, the bottom of the potential well corresponds to an equilibrium position at

$$a = a_{\text{eq}} \equiv -\frac{\beta}{\gamma E}. \tag{2.15}$$

Comparison with the expression (1.13) shows that the ansatz (2.6) with $a = a_{\text{eq}}$ exactly coincides with the unperturbed soliton solution.

In the case of constant β, Eq. (2.12) with the potential (2.13) is tantamount to the equation of motion for the radial variable r in the classical Kepler's problem (motion

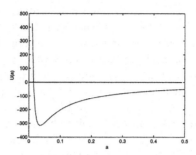

Figure 2.1: The effective potential (2.13) for $\beta = -1, \gamma E = 4\pi^2$.

of a particle in the gravitational field with the potential $-1/r$) [3]. Accordingly, exact solutions can be found in a parametric form,

$$a = -\left(\frac{2E}{\pi^2 H}\right)\left(1 - \sqrt{1 - \frac{\pi^2|H|}{2E^2}}\cos\xi\right), \quad \left(\frac{\pi^2|H|^{3/2}}{\sqrt{2}E}\right)z = \xi - \sqrt{1 - \frac{\pi^2|H|}{2E^2}}\sin\xi,$$

(2.16)

where it was set $-\beta = \gamma = 1$, ξ is an auxiliary dynamical variable, and H is the value of the Hamiltonian of the equivalent Kepler's problem, that takes values in the interval $-\left(2/\pi^2\right)E^2 < H < 0$ (the minimum value of H corresponds to the equilibrium position (2.15) at the well's bottom). The frequency of the oscillatory solution $a(z)$, as given by Eqs. (2.16) is

$$K = \frac{\pi^2|H|^{3/2}}{\sqrt{2}E}.$$

(2.17)

It takes the maximum value

$$K_0 = 2E^2/\pi$$

(2.18)

at $H = -\left(2/\pi^2\right)E^2$, which corresponds to small oscillations near the bottom of the potential well.

2.2.2   Soliton dynamics in the model with the harmonic modulation

Predictions of the variational approximation

As was mentioned above, in the case of the harmonic periodic modulation of the local GVD coefficient, such as in Eq. (2.5), one may expect resonances between intrinsic vibrations of the free soliton, which are described, within the framework of the VA, by the solution (2.16), and external modulation of the GVD coefficient. Possible resonances can be studied analytically in the case of shallow modulation, $\varepsilon \ll 1$, by expanding equation (2.12) around the equilibrium position (2.15), and retaining quadratic and cubic nonlinear terms in the expansion. The fundamental resonance corresponds to the case when the (spatial) frequency K_0 of free small oscillations near the equilibrium position, given by Eq. (2.18), is close to the spatial modulation frequency, which is 1 in Eq. (2.5). In other words, varying the initial soliton's energy E, one may expect that the *fundamental resonance* occurs in a vicinity of the value $E_{\text{fundam}} = \sqrt{\pi/2} \approx 1.25$. Further, the first *subharmonic resonance*, corresponding to $2K_0$ close to 1, and the *second-order resonance*, which takes place for K_0 close to 2, are expected around $E_{\text{subharm}} = \sqrt{\pi}/2 \approx 0.87$ and $E_{\text{second}} = \sqrt{\pi} \approx 1.77$, respectively.

Actual predictions of the VA should be obtained by from numerical simulations of Eq. ((2.12) with $\beta(z) = -(1 + \varepsilon \sin z)$. In particular, a solution with $a(z) \to \infty$ at $z \to \infty$ is interpreted as destruction of the soliton, as it becomes infinitely broad. In fact, this implies decay of the soliton into radiation.

The simulations performed in work [110] have demonstrated that, in an interval of values of the soliton's energies E which covers both the above-mentioned first subharmonic and second-order resonances, oscillations of $a(z)$ driven by the sinusoidal modulation of $\beta(z)$ are anharmonic but still periodic at very small values of ε, typically $\varepsilon \sim 0.01$. With the increase of the modulation depth ε, the oscillations become nonperiodic at $\varepsilon \sim 0.05$, and apparently chaotic at ε closer to 0.20. Finally, a critical value ε_{cr} can be found such that, at ε slightly larger than ε_{cr}, $a(z)$ performs a large number of irregular oscillations inside the potential wall, and then suddenly gets kicked out from the trapped state and escapes to infinity, as shown in a typical example in Fig. 2.2. The escape actually implies indefinite spreading out of the pulse, i.e., its eventual decay into radiation. In all the cases considered (with different values of the energy E), the critical modulation amplitude takes values in the interval

$$0.20 < \varepsilon_{\text{cr}} < 0.25 \,. \tag{2.19}$$

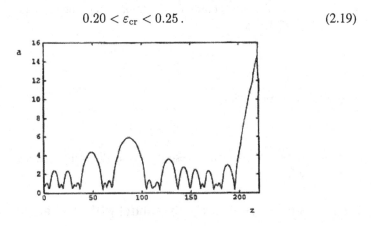

Figure 2.2: An example of the destruction of the soliton by a relatively weak harmonic modulation of the group-velocity-dispersion coefficient, $\beta(z) = -(1 + 0.25 \sin z)$, as predicted by numerical simulations of the variational equation (2.12). In this case, the energy is taken as $E = \sqrt{\pi} \approx 1.77$, which corresponds to the second-order resonance in small-amplitude driven oscillations near the bottom of the potential well (see the text). The width of the soliton, $a(z)$, performs a large number of irregular oscillations in the well, but finally escapes, which implies the decay of the soliton into radiation waves.

In the case when the modulation amplitude ε is small, the rate of direct emission of radiation by the soliton (obviously, this effect is beyond the scope of the VA) can be calculated by means of the perturbation theory [4]; however, this process does not play a crucial role in the destruction of the soliton.

Numerical results

The predictions produced by VA for the soliton in the NLS equation (2.5) with the sinusoidally modulated local GVD were compared with results of direct simulations of the equation in work [76]. Results of the simulations are summarized in Fig. 2.3. Two gross feature of this diagram roughly comply with predictions of the VA. Firstly, the destruction of the soliton may take place if the modulation amplitude exceeds a critical value, which varies, essentially, within an interval $0.15 < \varepsilon_{cr} < 0.20$, that should be compared to interval (2.19) predicted by VA. Secondly, the destruction of the soliton actually takes place, for ε not too large, if the initial squared soliton's energy E^2 exceeds a minimum value E^2_{min} varying between 1.8 and 2.0, which may be compared to the above-mentioned value $E^2_{fundam} = \pi/2$ that gives rise to the fundamental resonance between small vibrations of the perturbed soliton and the periodic modulation of the local GVD in Eq. (2.5).

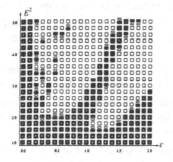

Figure 2.3: The phase diagram in the parametric plane (ε, E^2) for solitons in Eq. (2.5). The filled and unfilled rectangles correspond, respectively, to stable and splitting solitons.

The most essential difference between the assumptions on which the VA was based and numerical results is that the fundamental mode of the soliton destruction under the action of the sinusoidally modulated dispersion is not decay into radiation, but *splitting* of the soliton into two apparently stable secondary solitons (which is accompanied by emission of a considerable amount of radiation). A typical example of the splitting in displayed in Fig. 2.4. Obviously, the ansatz (2.6) does not admit any splitting; nevertheless, basic characteristics of the destruction of the soliton, even if the actual destructions mode is different from the one postulated by the VA, are predicted by the approximation qualitatively and semi-quantitatively correctly.

Detailed inspection of the numerical results shows that, prior to the splitting, the soliton performs a number of irregular vibrations, which resembles the picture produced by the VA, cf. Fig. 2.2. As well as in that picture, the vibrational stage preced-

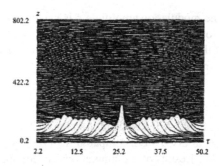

Figure 2.4: A typical example of the splitting of a fundamental soliton into two secondary ones in the NLS equation (2.5) with the sinusoidal modulation of the local
dispersion coefficient, for $\varepsilon = 0.3$ and $E^2 = 2.9$

ing the destruction of the soliton is quite long if the splitting takes place at ε that only
slightly exceeds the corresponding critical value $\varepsilon_{\mathrm{cr}}$.

The soliton stability diagram in the sinusoidally-modulated model (2.5), displayed
in Fig. 2.3, has a number of other noteworthy features, such as a narrow "stability
isthmus" , and a trend to restabilization of the soliton at large ε (note that for $\varepsilon > 1$,
the local GVD coefficient in Eq. (2.5) becomes sign-changing). These features are
observed in a parameter region where the VA cannot be used.

Essential additional results concerning the comparison between the VA and direct
simulations in the same model were reported, on the basis of direct simulations, by
Abdullaev and Caputo [3]. They have also found that the destruction of the soliton
takes place via its splitting into two secondary ones, and demonstrated that agreement
between the VA and direct simulations of small intrinsic vibrations of the soliton is
fairly good as long as the vibration frequency K_0, see Eq. (2.18), remains *smaller* than
the external modulation frequency (recall it is set to be 1 in Eq. (2.5)). At $K_0 \gtrsim 1$,
intensive emission of radiation takes place (even without complete destruction of the
soliton), which, naturally, strongly affects the agreement with the VA, as the approximation completely disregarded the radiation component of the field. Another important
numerical finding reported in work [3] is that, in cases when the variational and direct
numerical result are generally close, a more subtle effect of the radiation loss is strong
suppression of higher harmonics in the soliton's internal vibrations, in comparison with
the picture predicted by the VA.

2.3   Solitons in the model with dispersion management

This section deals with solitons in the practically important model of the long dispersion-
compensated nonlinear fiber link based on Eqs. (1.48) and (1.49). For convenience,
the equations are given here again:

$$iu_z - \frac{1}{2}\beta(z)u_{\tau\tau} + \gamma|u|^2u = 0, \qquad (2.20)$$

$$\beta = \begin{cases} \beta_1 > 0, & \text{if } 0 < z < L_1, \\ \beta_2 < 0, & \text{if } L_1 < z < L_1 + L_2 \equiv L, \end{cases} \tag{2.21}$$

The concept of the dispersion management (DM) for solitons in dispersion-compensated systems which, in the simplest case, amount to the model based on Eqs. (2.20) and (2.21), was introduced nearly simultaneously in works by Knox, Forysiak, and Doran [95], Suzuki, Morita, Edagawa, Yamamoto, Taga, and Akiba [160], Nakazawa and Kubota [131], and Gabitov and Turitsyn [69]. The original motivation for the development of the DM technique for solitons was the necessity to suppress the *Gordon-Haus effect*, i.e., random jitter of the soliton's center due to its interaction with optical noise, which is accumulated in the fiber-optic link due to the spontaneous emission from the optical amplifiers (these are actually Erbium-doped segments of the fiber, periodically inserted into the link, with the objective to compensate the fiber loss). The dispersion-managed solitons were predicted (and found in the experiment) to have large energy, which helps to improve the noise-to-signal ratio. Indeed, it has been demonstrated that the DM technique is very efficient in stabilizing solitons against the random jitter. On the other hand, a problem for the use of solitons in the DM system is posed by interactions between them. As the DM solitons periodically expand and contract, they may tangibly overlap, through their "tails", at the expansion stage, which leads to the increase of unwanted interaction between them. Besides their great significance to the applications, solitons in DM models have also drawn a great deal of attention as a subject of fundamental research. The account given below focuses chiefly on the latter point, although applied aspects are briefly considered too.

As explained below, the VA is a very natural tool to investigate the soliton dynamics in DM models. The presentation in this section will chiefly follow the approach elaborated in works [100] and [106] (the latter work applied the VA, in combination with direct simulations, to solitons in a model of *random DM*). In particular, the same normalization of parameters of the DM map (2.21) as in paper [100] is adopted here,

$$(\beta_1 - \beta_0) L_1 + (\beta_2 - \beta_0) L_2 = 0, |\beta_1 - \beta_0|L_1 = |\beta_2 - \beta_0|L_2 = 1, \ L \equiv L_1 + L_2 = 1, \tag{2.22}$$

which can be always imposed by means of an obvious rescaling (recall β_0 is the PAD defined as per Eq. (1.50)).

In the case of strong DM, when the nonlinear term in Eq. (2.20) may be treated as a small perturbation, the RZ Gaussian pulse (1.51), which is the exact solution in the linear limit, may be used as a natural variational ansatz, to take into regard effects of the weak nonlinearity. For convenience, the ansatz is written here again:

$$u_{\mathrm{RZ}}(z,\tau) = \sqrt{\frac{P_0}{1 - 2iB(z)/W_0^2}} \exp\left(-\frac{\tau^2}{W^2(z)} - \frac{2iB(z)}{W_0^2 W^2(z)}\right), \tag{2.23}$$

$$B(z) \equiv B_0 + \int_0^z \beta(z')dz' \tag{2.24}$$

The PAD will also be treated as a small perturbation, as an intuitive assumption is that the weak nonlinearity and small PAD may effectively compensate each other, supporting a robust RZ pulse (alias, DM soliton). An important dimensionless characteristic of the pulse (2.23) is its *dispersion-management strength*, which is defined as

$$S \equiv 1.443 \frac{|\beta_1| L_1 + |\beta_2| L_2}{W_0^2}. \tag{2.25}$$

The factor 1.443 appears here due to the use of the so-called FWHM (full-width-at-half-maximum) definition of the pulse's width, which is different from W_0. According to the value of S, the DM schemes are usually categorized as *weak-DM*, *moderate-DM*, and *strong-DM* regimes – in the cases of, roughly, $S < 3$, $S \sim 3 - 4$, and $S > 4$, respectively.

The VA is based on the assumption that the arbitrary constants W_0 and B_0 in Eqs. (2.23) and (2.24) become slowly varying functions of z, if the weak nonlinearity is taken into regard. Additionally, the accumulated dispersion $B(z)$ is defined for the map (2.21) from which the PAD value β_0 is subtracted. The analysis presented in detail in paper [100] yields the following evolution equations for the slowly varying parameters:

$$\frac{dW_0}{dz} = -\sqrt{2} \frac{E W_0 B(z)}{W^3(z)}, \tag{2.26}$$

$$\frac{dB_0}{dz} = \beta_0 - \frac{E \left[4B^2(z) - W_0^4 \right]}{2\sqrt{2}\, W^3(z)}, \tag{2.27}$$

where the function $W(z)$ is defined in Eq. (1.53). To derive these equations, the normalization (2.22) was used, and the energy of the pulse is defined as $E = P_0 \tau_0$ (recall that P_0 is peak power, i.e., maximum value of the squared amplitude, in expression (2.23)). In fact, E plays the role of a small parameter measuring the relative weakness of the nonlinearity in comparison with the local dispersion.

An issue of fundamental interest is to find conditions allowing for the *stationary transmission* of the pulse, i.e., a dynamical regime in which the parameters W_0 and B_0 return to their initial values after passing one DM period. Because, as it is seen from Eqs. (2.26) and (2.27), changes of W_0 and B_0 within one period are generally small, $\sim (\beta_0, E)$, in the first approximation one may insert unperturbed values of W_0 and B_0 into the right-hand sides of Eqs. (2.26) and (2.27), and impose the conditions that (recall that the DM period is 1 in the present notation)

$$\int_0^1 \frac{dW_0}{dz}\, dz = \int_0^1 \frac{dB_0}{dz}\, dz = 0. \tag{2.28}$$

After some analytical calculations, Eqs. (2.28) yield the following stationary-propagation conditions for the Gaussian pulse in an explicit form,

$$B_0 = \frac{1}{2},\ \beta_0 = \frac{\sqrt{2}}{4} E \tau_0^3 \left[\ln \left(\sqrt{1 + \frac{1}{W_0^4}} + \frac{1}{W_0^2} \right) - \frac{2}{\sqrt{W_0^4 + 1}} \right]. \tag{2.29}$$

The meaning of the condition $B_0 = 1/2$ is quite simple: it requires the pulse to have zero chirp at the midpoint of each fiber segment. The second condition (2.29) predicts that the DM soliton propagates steadily at anomalous PAD, $\beta_0 < 0$, provided that $W_0^2 > \left(W_0^2\right)_{\mathrm{cr}} \approx 0.30$, at $\beta_0 = 0$ if $W_0^2 = \left(W_0^2\right)_{\mathrm{cr}}$, and at *normal* PAD, $\beta_0 > 0$, if $\left(W_0^2\right)_{\mathrm{min}} \approx 0.148 < W_0^2 < \left(W_0^2\right)_{\mathrm{cr}}$. The latter case is very interesting, as the classical NLS soliton cannot exist at normal dispersion. Further analysis of Eq. (2.29) shows that, in this case, the solution exists in a limited interval of the normal-PAD values,

$$0 \le \beta_0/E \le (\beta_0/E)_{\mathrm{max}} \approx 0.0127. \tag{2.30}$$

Inside this interval, Eq. (2.29) yields *two* different values of the minimum width W_0 for a given value of β_0/E, while in the anomalous-PAD region, W_0 is a uniquely defined function of β_0/E. In the case of the normal PAD, $\beta_0 > 0$, it can be concluded that the DM soliton corresponding to the larger value of W_0 is stable, while the one corresponding to smaller W_0 is unstable. The border between the stable and unstable solitons corresponds to $\beta_0/E = (\beta_0/E)_{\mathrm{max}}$ (see Eq. (2.30)), and it is located at $W_0^2 = \left(W_0^2\right)_{\mathrm{min}} \approx 0.148$ (i.e., all the stable and unstable solitons have, respectively, $W_0^2 > \left(W_0^2\right)_{\mathrm{min}}$ and $W_0^2 < \left(W_0^2\right)_{\mathrm{min}}$). The results concerning the stability of these two solitons were substantiated in a mathematically rigorous form in work [140].

Translating W_0^2 into the standard DM-strength parameter S according to Eq. (2.25), one concludes that the VA predicts the following:

- stable DM solitons at anomalous PAD if $S < S_{\mathrm{cr}} \approx 4.79$;

- stable DM solitons at *zero* PAD if $S = S_{\mathrm{cr}} \approx 4.79$;

- stable DM solitons at *normal* PAD if $4.79 < S < S_{\mathrm{max}} \approx 9.75$;

- no stable DM soliton for $S > S_{\mathrm{max}} \approx 9.75$.

The normalized power of the DM soliton, which is $P \equiv 4 \cdot 1.12 P_0$ (the factor 1.12 is the ratio of the FWHM widths for the sech-shaped and Gaussian pulses) is shown vs. the DM strength, at several fixed values of PAD β_0, as predicted by Eq. (2.29), in Fig. 2.5. A counterpart of the same dependence, obtained in work [31] from direct simulations of the underlying equation (2.20), is displayed in Fig. 2.6 (a typical example of the shape of the DM soliton, found from direct simulations, is displayed below in Fig. 2.10). In Fig. 2.5, the curves are shown only in the region of $S < 9.75$, where the solitons are expected to be stable. The curves in Fig. 2.6 corresponding to the normal PAD ($\beta_0 > 0$) terminate at points where the corresponding DM soliton becomes unstable.

As a matter of fact, Fig. 2.6 is the most fundamental and comprehensive character-istic of the family of the DM solitons. The comparison of Figs. 2.5 and 2.6 shows that the VA yields acceptable results for relatively small values of the soliton's power, where the underlying assumption, that the nonlinearity may be treated as a weak perturbation, is relevant. In particular, the VA-predicted critical value $S_{\mathrm{cr}} \approx 4.79$ is different from but nevertheless reasonably close to the critical DM strength $S_{\mathrm{cr}} \approx 4$ which direct sim-ulations yield in the small-power limit. With the increase of power, numerically found

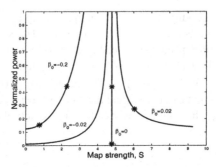

Figure 2.5: The peak power of the stable soliton in the DM system vs. the map strength S at different values of the path-average dispersion β_0, as predicted by the variational approximation based on Eq. (2.29). Here and in the next figure, stars mark cases for which the corresponding model with *random DM* was investigated in detail, see section 2.4.

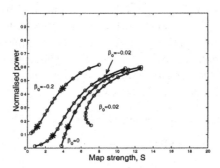

Figure 2.6: A counterpart of Fig. 2.5 obtained from direct numerical simulations of Eq. (2.20).

S_{cr} grows. It is also noteworthy that the value $S_{\max} \approx 9.75$, predicted by the VA as the stability limit for the DM solitons, is indeed close to what is revealed by direct simulations for small powers, as seen in Fig. 2.6.

The DM soliton considered above is a fundamental one (i.e., it always keeps the single-peak shape). Higher-order DM solitons can be constructed too. Indeed, along with the expression (2.23), its τ-derivatives of all orders are also exact solutions to the linearized version of Eq. (2.20), and can be used as *ansätze* to generate an (approximate) higher-order soliton solutions in the weakly nonlinear case. In particular, an ansatz proportional to the first derivative of the Gaussian (2.23),

$$\left(u_{\mathrm{RZ}}^{(\mathrm{linear})}(z,\tau)\right)_{\mathrm{odd}} = \sqrt{\frac{P_0}{1 - 2iB(z)/W_0^2}} \frac{\tau}{W^2(z)} \exp\left(-\frac{\tau^2}{W^2(z)} - \frac{2iB(z)}{W_0^2 W^2(z)}\right),$$

$$(2.31)$$

was used in work [137] to construct an odd (antisymmetric, as a function of τ) DM soliton. However, this soliton is unstable against even perturbations. Related to this generalization is another technical approach to the description of the fundamental DM solitons and perturbations around them: an extended solution, including the perturbation, may be looked for starting from a linear combination of Hermite-Gauss functions (of τ) [167, 99]. In particular, this approach correctly reproduces the results of the VA.

2.4 Random dispersion management

Existing terrestrial fiber-optic telecommunication networks are patchwork systems, which include links with very different lengths [16]. This practically important circumstance suggests to consider transmission of RZ pulses (quasi-solitons) in *random* DM systems. It was shown that VA applies to this case too [106]. Random-DM models of different types were considered in works [1] and [73], where local values of the GVD coefficient, rather than the fiber-segment lengths, are distributed randomly. Actually, search for robust RZ pulses in random nonlinear fiber-optic systems is an issue not only important to the applications, but also fundamentally significant to the general theory of nonlinear waves in disordered media [96].

The basic equation and normalizations for the DM system with random distribution of the cell lengths can be taken in the same form as in the previous section, i.e., as per Eqs. (2.20), (2.21), and (2.22), with a difference that, in the random system, the normalizations must be applied to mean values of the randomly varying lengths. The consideration is limited here to the most important case when the lengths of the segments with the anomalous and normal GVD are equal in each DM cell, $L_1 = L_2 \equiv L/2$. Then, Eqs. (2.22) yield the mean values $\overline{L}_{1,2} = 1/2$, and $|\beta_{1,2} - \beta_0| = 2$. To comply with the former condition, one may assume that the random lengths are distributed uniformly in the interval

$$0.1 < L/2 < 0.9 . \tag{2.32}$$

The minimum length 0.1 is introduced because, in reality, the length can be neither very large (say, larger than 200 km) nor very small (shorter than 20 km).

The same ansatz (2.23) and variational equations (2.26) and (2.27) which was applied above to the regular (periodic) DM system, may be used with its random counterpart. As explained above, the change of the soliton's parameters, $W_0 \rightarrow W_0 + \delta W_0$, $B_0 \rightarrow B_0 + \delta B_0$, within one DM cell is small. Therefore, the evolution of the pulse passing many cells may be approximated by smoothed differential equations,

$$\frac{dW_0}{dz} = \frac{\delta W_0}{L^{(n)}}, \frac{dB_0}{dz} = \frac{\delta B_0}{L^{(n)}} \tag{2.33}$$

(n is the cell's number). Finally, the equations take the following form,

$$\frac{dW_0}{dz} = \frac{\sqrt{2}EW_0^4}{4L(z)}\left[\frac{1}{\sqrt{W_0^4 + 4B_0^2}} - \frac{1}{\sqrt{W_0^4 + 4\left(B_0 - L(z)\right)^2}}\right], \quad (2.34)$$

$$\frac{dB_0}{dz} = B_0 + \frac{\sqrt{2}EW_0^3}{8L(z)}\left[\frac{4B_0}{\sqrt{W_0^4 + 4B_0^2}} - \frac{4\left(B_0 - L\right)}{\sqrt{W_0^4 + 4\left(B_0 - L(z)\right)^2}}\right.$$

$$\left. + \ln\left(\frac{\sqrt{W_0^4 + 4B_0^2} - 2B_0}{\sqrt{W_0^4 + 4\left(B_0 - L(z)\right)^2} - 2\left(B_0 - L(z)\right)}\right)\right], \quad (2.35)$$

where $L(z)$ is regarded as a continuous random function with values uniformly distributed in the interval (2.32).

The most essential characteristic of the pulse propagation at given values of β_0 and E is the cell-average pulse's width,

$$\bar{W} \equiv L^{-1}\int_{\text{cell}} W(z)dz. \quad (2.36)$$

Simulations of Eqs. (2.34) and (2.35) reveal that there are two drastically different dynamical regimes. If the soliton's energy is sufficiently small (hence the approximation outlined in the previous section is relevant) and PAD is anomalous or zero, i.e., $\beta_0 \leq 0$ (especially, if $\beta_0 = 0$), the pulse performs random vibrations but remains truly stable over long propagation distances. In the case when the energy is larger, as well as when PAD is normal, $\beta_0 > 0$, the pulse suffers fast degradation.

Following work [106], typical examples of the propagation are displayed in Fig. 2.7 for the zero-PAD case, which is the best one in terms of the soliton stability. Simulations of Eqs. (2.34) and (2.35) were performed with 200 different realizations of the random function $L(z)$. Figure 2.7 displays the evolution of $\langle \bar{W}(z)\rangle$, i.e., mean value of the width (2.36) averaged over the 200 random realizations, along with the corresponding normal deviations from the mean value. The figure demonstrates that some systematic slow evolution takes place on top of the random vibrations, which are eliminated by averaging over 200 realizations. Systematic degradation (broadening) of the soliton takes place too, but it is extremely slow if the energy is small. In the case shown in the bottom part of Fig. 2.7, the pulse survives, with very little degradation, the transmission through more than 1000 average cell lengths (in fact, as long as the simulations could be run). It is not difficult to understand this: in the limit of zero power, i.e., in the linear random-DM model, an exact solution for the pulse is available in an essentially the same form as given above for the periodic DM, see Eq. (2.23). If PAD is exactly zero, this exact solution predicts no systematic broadening of the pulse.

If the soliton's energy is larger, further simulations of Eqs. (2.34) and (2.35) show that, after having passed a very large distance, the slow spreading out of the soliton suddenly ends up with its blowup (complete decay into radiation). This seems to be qualitatively similar to what was predicted by the VA in the case of the periodic sinusoidal modulation of the dispersion, as shown in Fig. 2.2: a long sequence of chaotic but nevertheless quasi-stable vibrations is suddenly changed by rapid irreversible decay.

In fact, the case of $\beta_0 = 0$ is a point of a *sharp optimum* for the random-DM system: at any finite anomalous PAD, $\beta_0 < 0$, the degradation of the pulses is essentially faster, especially for pulses with larger energy, and at any small normal value of PAD, $\beta_0 > 0$, very rapid decay always takes place, virtually at all values of the energy.

Comparison of the results predicted by VA with direct simulations of the full random-DM model was also reported in work [106]. Direct numerical results, averaged over the same number (200) of the realizations of the random-length set $L^{(n)}$, turn out to be quite similar to what was predicted by VA. In particular, the most stable propagation is again observed at zero PAD, the soliton's broadening is faster at nonzero anomalous PAD, and all solitons decay very quickly at nonzero normal PAD. The soliton's stability in direct simulations drastically deteriorates with the increase of the energy, as was also predicted by the VA.

Detailed comparison shows that, surprisingly, direct simulations yield somewhat *better* results for the soliton's stability than the VA: the actual broadening rate of the soliton may be $\sim 20\%$ smaller than that predicted by VA. The slow long-scale oscillations, clearly seen in Fig. 2.7, are less pronounced in direct simulations. The sudden decay into radiation, predicted by the VA after very long propagation, is not observed in direct simulations; instead, the soliton eventually splits into two smaller ones, quite similar to what is observed in direct simulations of the model with the periodically modulated dispersion, see Fig. 2.4.

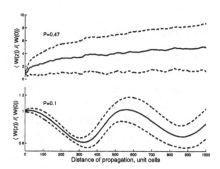

Figure 2.7: Evolution of the cell-average pulse width in the random-DM system with zero path-average dispersion, as predicted by the variational approximation. The propagation distance exceeds 1000 DM cells. The top and bottom panels correspond, respectively, to high and low power, $P = 0.47$ and $P = 0.1$. The mean values (solid curve) and standard deviations from them (dashed curves) are produced by numerical integration of equations (2.34) and (2.35), followed by averaging over 200 different realizations of the random length set.

To conclude this section, it is relevant to mention that some other versions of the DM systems, also different from strictly periodic ones, were studied too. In particular, an interesting possibility is to consider the so-called "hyperbolic" model, in which the size of the DM cell is fixed, while the PAD gradually decreases, as $1/z$, with the propagation distance (which is achieved by a slow systematic change of the ratio of the anomalous- and normal-GVD segments in the DM map). It was demonstrated that the system of the latter type is especially efficient in suppressing the soliton's jitter [178].

2.5 Dispersion-managed solitons in the system with loss, gain and filtering

As mentioned above, a strong incentive for the introduction of the DM schemes for solitons was the potential that they offer for suppression of the jitter induced by the optical noise in the fiber through the Gordon-Haus effect. However, the DM alone cannot provide for complete suppression of the jitter, that is why long-haul dispersion-managed links for the transmission of solitons must include optical bandpass filters [120] (the filters are well known to be a versatile tool of the jitter control [16]). Therefore, it is necessary to modify the theory outlined above, in order to take the filtering into regard. Simultaneously, a model of the real-world fiber-optic link must include fiber loss and compensating gain provided by linear amplifiers periodically inserted into the link (these very amplifiers are also the main source of the optical noise that gives rise to the jitter). In this section, basic results for solitons in the filtered DM model will be presented, following the analysis developed in works [32] and [40]. A qualitative difference from the results outlined above for the lossless model is that there is a minimum value of the pulse's peak power necessary for the existence of stable solitons.

2.5.1 Distributed-filtering approximation

In the most typical case, the characteristic soliton period z_{sol} in long-haul fiber links (see Eq. (1.16) for the definition of z_{sol}) is quite large, $\sim 200 - 300$ km, while the spacing between the amplifiers is essentially smaller, $\sim 50 - 80$ km. Normally, bandpass filters are integrated with amplifiers, therefore the spacing between the filters is relatively small too. This suggests to adopt the approximation in which the corresponding modified NLS equation (2.20) neglects the discrete placing of the amplifiers and filters, assuming that they are distributed uniformly along the link. The accordingly modified NLS equation takes the form

$$iu_z - \frac{1}{2}\beta(z)u_{\tau\tau} + \gamma|u|^2 u = ig_0 u + ig_1 u_{\tau\tau}, \qquad (2.37)$$

where $g_1 > 0$ is the effective filtering coefficient, and $g_0 > 0$ is gain necessary to compensate the filtering losses; the model implies that the fiber loss proper is compensated by the main part of the gain, that does not appear in Eq. (2.37) in the distributed-gain approximation. It is necessary to stress that, despite the use of the assumption postulating the uniformly distributed filtering and gain, the model based on Eq. (2.37) belongs to the general class of the periodic strongly inhomogeneous nonlinear systems, due to the presence of the DM on the left-hand side of the equation.

Following the lines of the analysis developed for the lossless model, one can treat the filtering and gain terms in Eq. (2.37) as additional small perturbations (which completely corresponds to the realistic conditions in fiber-optic telecommunications). The perturbation theory may again be based on the ansatz (2.23), which, by itself, is the exact solution of Eq. (2.20) in the absence of the nonlinearity, filtering and gain. One can easily see that, in fact, the Gaussian ansatz provides for an *exact solution* to Eq. (2.37) if the nonlinearity is still neglected, while the linear terms on the right-hand side are taken into regard. The corresponding exact solution is obtained from (2.23) by the substitution

$$B(z) \to \tilde{B}(z) \equiv \int_0^z \beta(z')\,dz' + B_0 - ig_1 z, \quad P_0 \to P_0 e^{2g_0 z} \qquad (2.38)$$

(note that the accumulated dispersion $\tilde{B}$, modified by the filtering, is complex).

Next, one can analyze conditions which single out established DM solitons, by demanding zero changes of parameters W_0 and B_0 after the passage of one DM map (recall these conditions result in Eqs. (2.29) in the lossless model). As follows from Eq. (2.38) and from the way the accumulated dispersion enters the Gaussian ansatz (2.23), the filtering and gain do not affect the evolution of B_0, and generate an additional small change,

$$\Delta W_0 = 2g_1 L/W_0, \qquad (2.39)$$

of the width parameter of the pulse passing the distance L corresponding to the DM map (the meaning of this result is quite simple: the filtering gives rise to spreading out of the pulse at a constant rate). A new condition, which is enforced by the filtering and gain terms, is that the energy of the pulse must also remain equal to the initial value after the passage of the DM map (in the conservative model, this condition holds automatically, provided that emission of linear "radiation" from the soliton may be neglected). To realize this additional condition, one should notice that the terms on the right-hand side of Eq. (2.37) give rise to the following exact evolution equation for the soliton's energy E (which is defined as per Eq. (1.9)):

$$\frac{dE}{dz} \equiv \frac{d}{dz}\left(\frac{1}{2}\int_{-\infty}^{+\infty}|u(\tau)|^2\,d\tau\right) = g_0\int_{-\infty}^{+\infty}|u(\tau)|^2\,d\tau - g_1\int_{-\infty}^{+\infty}|u_\tau|^2\,d\tau.$$
$$(2.40)$$

The substitution of the ansatz (2.23) into this equation and calculation of the integrals yields an explicit result for dE/dz, which can be further integrated over the interval $\Delta z = L$ corresponding to the DM-cell's length. Finally, equating the energy change to zero yields a very simple relation which shows that the balance between the filtering loss and compensating gain uniquely selects the width parameter W_0,

$$W_0^2 = g_1/g_0 \qquad (2.41)$$

(which remained arbitrary without the filters [100]). Actually, Eq. (2.41) may be understood in a different way: for given W_0, it determines the necessary value g_0 of the gain.

Amending the second relation in Eqs. (2.29) with regard to the filtering-induced change (2.39) of the width, and defining, for convenience, $\kappa \equiv \sqrt{2}/W_0$ and $\phi =$

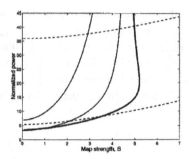

Figure 2.8: The diagram of the stationary transmission regimes for soliton in the DM model with filtering, as obtained from the analytically derived equations (2.42) and (2.43). The thin solid, bold solid, and dashed curves are reference lines which correspond, respectively, to the values $\beta_0/g_1 = -5$ and -1 (anomalous dispersion), $\beta_0/g_1 = 0$, and $\beta_0/g_1 = 1$ and 5 (normal dispersion).

$-2B_0/W_0^2$, the conditions for the pulse to be stationary are (the same normalizations (2.22) as in the lossless model are implied here)

$$\frac{\gamma P_0}{g_1}\left(\frac{1}{\sqrt{1+\phi^2}} - \frac{1}{\sqrt{1+(\phi+\kappa^2)^2}}\right) + \sqrt{2}\kappa^4 = 0, \tag{2.42}$$

$$\frac{\gamma P_0}{g_1}\left(\ln\left[\frac{\phi+\kappa^2+\sqrt{1+(\phi+\kappa^2)^2}}{\phi+\sqrt{1+\phi^2}}\right] - \frac{2\kappa^2}{\sqrt{1+(\phi+\kappa^2)^2}}\right) - \sqrt{2}\left(\frac{\beta_0}{g_1} - 2\phi\right)\kappa^4 = 0. \tag{2.43}$$

(recall γ is the constant nonlinearity coefficient in Eq. (2.37)).

By solving Eqs. (2.42) and (2.43), one can find the dependence of the pulse's peak power P_0 on the pulse's width and *normalized* PAD, β_0/g_1, in the presence of the filters. To facilitate the comparison with the lossless model (see Figs. 2.5), (2.6)), the results are displayed in Fig. 2.8 in terms of the map strength (2.25) and normalized peak power. Following work [32], the latter is taken as $0.22\gamma P_0 W_0^2/g_1$.

The results predicted by the analytical equations (2.42) and (2.43) are compared with results of direct numerical simulations in Fig. 2.9. Strictly speaking, the pulses in Eq. (2.37) are unstable because the gain term makes the trivial solution, $u = 0$, i.e., the background on which the soliton sits, unstable. Indeed, for high values of the gain, no stable solutions could be found numerically. Nevertheless, it is easy to find pulses that remain completely stable, at values of the parameters used in Fig. 2.9, after having passed, at least, 100 DM-map lengths, which is more than sufficient for the applications.

It is seen from the comparison of Figs. 2.8 and 2.9 that the analytical and numerical findings are in qualitative agreement. Further, comparing these figures to Figs. 2.5 and 2.6 which display analogous results for the lossless model, one notices a principally novel feature: the critical strength, $S_{cr} \approx 4$, which separates the stable transmission of

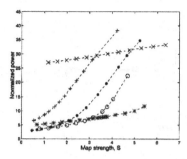

Figure 2.9: A counterpart of the diagram of stable soliton states shown in Fig. (2.8), as obtained from direct numerical simulations of Eq. (2.37), for $\beta_0/g_1 = -5$ ('+'), -1 (dot), 0 ('o'), 1 ('*') and 5 ('x').

the DM solitons at the anomalous and normal PAD, is removed, the stable transmission at zero and normal PAD being now possible at *any* map strength S. Instead, there appears, for the fixed filtering strength g_1, a minimum (critical) power which is necessary for the stable transmission of the RZ pulses (solitons) in the filtered DM system. In particular, it was shown in work [32] that an absolute minimum of the normalized power, $\simeq 3.3$, is found for $S = 0$ and weakly anomalous PAD, $\beta_0/g_1 \simeq -0.7$.

The general conclusion (which is supported by more detailed numerical results [32]) is that the filtering makes the DM solitons essentially less sensitive to the exact value of PAD. This feature can be quite beneficial for the applications. In particular, in a multi-channel (WDM) system, the PAD may alter from a channel to a channel, because of the presence of the third-order dispersion in the fiber. The filtering will make the system more stable not only against the Gordon-Haus jitter, but also against the scatter of the PAD values.

2.5.2 The lumped-filtering system

The results obtained in the approximation of the uniformly distributed filtering and gain generally correctly describe qualitative properties of realistic systems with *lumped* (discretely placed) filters and amplifiers. Nevertheless, important features are missed by the distributed-filtering approximation – in particular, specific instability occurs if the filters are placed at "wrong" positions relative to the DM map, namely, at midpoints of the normal-GVD segments (while the transmission of the solitons is completely stable with the filters set at midpoints of the anomalous-DM segments) [121].

A full stability diagram for the DM solitons in the model with lumped amplifiers was recently obtained in work [40]. The model is based on Eq. (2.37), with the right-hand side replaced by the following lumped-filtering expression:

$$ig_0 u + ig_1 u_{\tau\tau} \rightarrow i\sum_n \delta\left(z - z_0 - Ln\right)\left(g_0 u + \hat{G}u\right), \qquad (2.44)$$

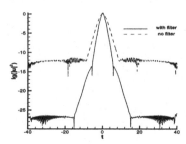

Figure 2.10: A typical example of the shape (shown on the logarithmic scale) of dispersion-managed solitons in the ideal lossless model, and in its realistic counterpart with lumped filters and amplifiers (for comparison, the solitons with equal amplitudes are taken in both cases). Each soliton is shown at a position (close to the midpoint of the anomalous-GVD segment of the DM map) where it is narrowest.

where the filtering operator $\hat{G}$ is defined by its action on a temporal harmonic,

$$\hat{G}\left(e^{-i\omega\tau}\right) = -\left(\frac{\omega}{\Delta\omega}\right)e^{-i\omega\tau}, \ \Delta\omega = \text{const} \qquad (2.45)$$

(the latter is usually referred to as the Gaussian transfer function with the bandpass width $\Delta\omega$). A typical example of the shape of a stable soliton in this realistic model is shown (on the logarithmic scale of the power, which is relevant to display the soliton's shape) in Fig. 2.10; for comparison, the shape of the soliton in the lossless model, with the same DM map and the same value of the soliton's peak power, is also displayed in the figure. As is seen, a beneficial effect produced by the filters is suppression of the soliton's "wings", which is quite important to attenuate unwanted interactions between solitons carrying the data stream in a fiber-optic telecommunication link.

A full stability diagram for the DM solitons in this model was generated by means of numerical methods specially developed for this purpose in work [40]. The diagram is displayed in Fig. 2.11 (the stability borders shown in this figure are, actually, some-what fuzzy, as, close to the borders, the RZ pulses observed in numerical simulations are still stable – in the sense that they do not decay – but they demonstrate irregular oscillations, and also change the shape, developing relatively large sidelobes attached to the main body of the soliton). Comparison of Figs. 2.11 and 2.9 shows that the distributed-filtering model indeed provides for a generally reasonable approximation to the lumped-filtering system.

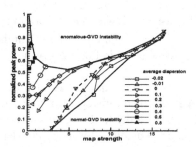

Figure 2.11: Stability diagram for the DM solitons in the realistic model with lumped filtering and amplification, given by Eqs. (2.37), (2.44), and (2.45). Symbol chains show stable solitons with different fixed values of the path-average dispersion (mean GVD). The overall stability area is bordered by the bold curves.

2.6 Collisions between solitons, and bound states of solitons in a two-channel dispersion-managed system

2.6.1 Effects of inter-channel collisions

The wavelength-division multiplexing (WDM), i.e., the use of a large number of data-bearing channels in the same fiber, carried by different wavelengths, is the most important direction in the development of optical telecommunications. In soliton-based systems, a serious problems posed by WDM is *crosstalk* due to collisions of pulses belonging to different channels. Collisions are inevitable, as the fiber's GVD gives rise to a difference in the group velocity (GVM) between the carrier waves in different channels.

In addition to the above-mentioned benefits of the jitter suppression in DM systems, another strong insensitive to study the soliton dynamics in DM systems is the fact the DM provides for strong suppression of the inter-channel crosstalk, as was first shown in direct simulations of a two-channel DM model reported in paper [134]. Besides that, collisions between solitons in coupled channels is a subject of considerable interest from the viewpoint of the general theory of solitons in periodic strongly inhomogeneous systems. In this section, an account of the analysis of collisions in the two-channel DM system will be given, following, chiefly, work [88].

The simplest two-channel system is described by two NLS equations coupled by the XPM (cross-phase-modulation) terms (cf. Eqs. (1.22) and (1.23)),

$$i\left(u_z + cu_\tau\right) - \frac{1}{2}\beta(z)u_{\tau\tau} + \left[-\frac{1}{2}\beta_0^{(u)}u_{\tau\tau} + \gamma\left(|v|^2 + 2|u|^2\right)u\right] = 0, \quad (2.46)$$

$$i\left(v_z - cv_\tau\right) - \frac{1}{2}\beta(z)v_{\tau\tau} + \left[-\frac{1}{2}\beta_0^{(v)}v_{\tau\tau} + \gamma\left(|u|^2 + 2|v|^2\right)v\right] = 0, \quad (2.47)$$

where $2c$ is the inverse-group-velocity difference between the channels (i.e., the GVM), $\beta(z)$ is the main part of the dispersion, with the zero average, that may be assumed the same in both channels, $\beta_0^{(u,v)}$ are values of the PAD in the channels, which are, generally, different. The nonlinear terms in Eqs. (2.46) and (2.47) represent the SPM and XPM effects.

An analytical approach to the problem may be based on the ansatz of the type (2.23) for the soliton in each channel, modified with regard to the inverse-group-velocity shifts $\pm c$ in Eqs. (2.46) and (2.47). In order to describe the dynamics of the interacting pulses, the ansatz should be further modified by applying independent Galilean boosts to u and v (cf. the boost formula (1.6) for $\beta = $ const):

$$
\begin{aligned}
u(z,\tau;\omega) &= u_{\mathrm{RZ}}^{(\mathrm{linear})}\left(z,\tau - cz - T_u(z)\right)\exp\left[-i\omega_u\tau + i\psi_u(z)\right], \\
v(z,\tau;\omega) &= u_{\mathrm{RZ}}^{(\mathrm{linear})}\left(z,\tau + cz - T_v(z)\right)\exp\left[-i\omega_v\tau + i\psi_v(z)\right].
\end{aligned} \quad (2.48)
$$

Here, ω_u and ω_v are frequency shifts of the two solitons, and the corresponding position shifts obey the equations

$$
\frac{dT_u}{dz} = \omega_u\left[\beta(z) + \beta_0^{(u)}\right], \quad \frac{dT_v}{dz} = \omega_v\left[\beta(z) + \beta_0^{(v)}\right]. \quad (2.49)
$$

In the absence of the interaction, parameters of the DM solitons in both channels are selected by the conditions (2.29). Since these conditions were obtained by treating the SPM nonlinearity as a small perturbation, the XPM-induced interaction between solitons may also be considered as another perturbation. This approach makes it possible to derive (by means of the Lagrangian technique, as shown in detail in work [88]) the following XPM-induced evolution equations for the frequency shifts of the solitons in the u- and v-channels:

$$
\frac{d}{dz}\left\{\begin{matrix}\omega_u \\ \omega_v\end{matrix}\right\} = \frac{2^{5/2}P_{v,u}W_0^4 cz}{\left[W_0^4 + 4B^2(z)\right]^{3/2}}\left\{\begin{matrix}+P_v \\ -P_u\end{matrix}\right\}\exp\left[-\frac{W_0^2\left(\Delta T(z)\right)^2}{W_0^4 + 4B^2(z)}\right], \quad (2.50)
$$

where P_u and P_v are the peak powers of the pulse in the u- and v-channels, and $\Delta T(z)$ is the temporal separation between the solitons. According to *ansätze* (2.48) and Eqs. (2.49), $\Delta T(z)$ obeys the equation

$$
\frac{d}{dz}\Delta T = 2c + \omega_u\left[\beta(z) + \beta_0^{(u)}\right] - \omega_v\left[\beta(z) + \beta_0^{(v)}\right]. \quad (2.51)
$$

The dynamical equations (2.50) are not only important for the application to optical telecommunications, but also help to understand the nature of the soliton's dynamics in the system: in the present approximation, the solitons may be regarded as quasi-particles with the coordinates T_u and T_v, if the evolution variable z is treated as formal time, $\omega_u + c$ and $\omega_v - c$ being momenta of the particles. In terms of this mechanical interpretation, $\beta(z) + \beta_0^{(u)}$ and $\beta(z) + \beta_0^{(v)}$ are proportional to the corresponding inverse masses (note that the DM solitons are characterized by the *time-dependent* effective masses, which periodically flip the sign), and Eq. (2.50) is a force of the interaction between the particles, which depends on the distance ΔT between them. In this connection, it is relevant to mention that a dynamical equation similar to (2.50) was derived in work [10] for a two-channel model with constant GVD. However, there is a principal difference between the collisions in the two-channel DM system and in its constant-dispersion counterpart: as the coefficient $\beta(z)$ in Eqs. (2.49) periodically changes its sign (i.e., the effective masses periodically flip their sign, as mentioned above), in the strong-DM regime colliding pulses pass through each other *many times* before separating.

In the case relevant for the application to the WDM system in optical telecommunications, the term $2c$ is Eq. (2.51) is much larger than the two other terms [88], therefore Eq. (2.50) may be replaced by a simpler one, with

$$\Delta T(z) = 2cz \qquad (2.52)$$

(however, a case is also possible when this assumption is not valid; then, the two solitons may form a *bound state* [63], see below).

It is necessary to distinguish between *complete* and *incomplete* collisions. In the former case, the solitons are far separated before the collision, while in the latter case, which corresponds to a situation when the collision occurs close to the input point ($z = 0$), the solitons begin the interaction being strongly overlapped. In either case, the most important result of the collision are shifts of the soliton's frequencies $\delta\omega_u$ and $\delta\omega_v$, which can be calculated as

$$\delta\omega_{u,v} = \int_{z_0}^{+\infty} \frac{d\omega_{u,v}}{dz} dz. \qquad (2.53)$$

Here $d\omega_{u,v}/dz$ are to be taken from Eq. (2.50), with $\Delta T(z)$ replaced by $2cz$, as per Eq. (2.52). The lower limit of the integration in the expression (2.53) is finite for the incomplete collision, while the complete collision corresponds to $z_0 = -\infty$. The frequency shift is very detrimental in terms of the fiber-optic telecommunications, as, through the GVD, it gives rise to a change of the soliton's velocity. If the soliton picks up a "wrong" velocity, information carried by the soliton stream in the fiber-optic link may be completely lost.

An estimate of physical parameters for the *dense* WDM arrangements, with the actual wavelength separation between the channels $\delta\lambda < 1$ nm (this is the case of paramount practical interest) shows that, although the term $2c$ dominates in Eq. (2.51), c may be regarded as a small parameter in the integral expression (2.53), in the sense that the function cz varies slowly in comparison with the rapidly oscillating accumulated dispersion $B(z)$. In this case, the integral (2.53) and similar integrals can be calculated in a fully analytical form, as shown in work [88]. In particular, Eq. (2.53) yields *zero* net frequency shift for the complete collision, which shows the ability of the DM to suppress collision-induced effects. In fact, the zero net shift is a result of the multiple character of the collision (see above): each elementary collision generates a finite frequency shift, but they sum up to zero.

If the frequency shift is zero, the collision is characterized by a position shift, which is a detrimental effect too in optical telecommunications, but less dangerous than the frequency shift. The position shift can be found from Eq. (2.49),

$$\delta T_{u,v} \equiv \int_{-\infty}^{+\infty} \frac{dT_{u,v}}{dz} dz = -\epsilon \beta_0^{(u,v)} \int_{-\infty}^{+\infty} z \frac{d\omega_{u,v}}{dz} dz - \int_{-\infty}^{+\infty} B(z) \frac{d\omega_{u,v}}{dz} dz,$$

(2.54)

where integration by parts was done. Then, substituting the expression (2.50) for $d\omega_u/dz$, one can perform the integrations analytically, to obtain a very simple final result:

$$\left\{ \begin{array}{c} \delta T_u \\ \delta T_v \end{array} \right\} = \frac{\sqrt{2\pi} W_0}{4c^2} \left\{ \begin{array}{c} \beta_0^{(u)} P_v \\ \beta_0^{(u)} P_u \end{array} \right\}.$$

(2.55)

This result contains a product of two small parameters, namely, the PAD $\beta_0^{(u,v)}$ and power $P_{v,u}$ (the latter is small as it measures the nonlinearity in the system, and it was assumed from the very beginning that the nonlinearity is a small perturbation).

The frequency shift generated by the incomplete collision can be found similarly. In this case, the worst (largest) result is obtained for the configuration with centers of the two solitons coinciding at the launching point $z = 0$:

$$(\delta\omega_{u,v})_{\max} = \frac{P_{v,u}}{\sqrt{2}cS} \ln\left(S + \sqrt{1 + S^2}\right),$$

(2.56)

where S is the DM strength defined by Eq. (2.25).

These analytical results were compared with numerical simulations. First of all, simulations show that the frequency shift generated by complete collisions is very small indeed (much smaller than in the case of incomplete collisions at the same values of parameters). As for the position shift in the case of the complete collision, the analytical prediction (2.55) for it is compared to numerical results in Fig. 2.12, showing a reasonably good agreement. In the case of incomplete collisions, simulations yield a nonzero frequency shift, which was compared to the analytical prediction (2.56) in work [88], also showing a reasonable agreement.

2.6.2 Inter-channel bound states

The case when the last two terms in Eq. (2.51) are comparable to $2c$ is relatively exotic but physically possible too. For this case, formation of *stable bound states* of two solitons belonging to the different channels was predicted in work [63]. In physical units, the "exotic" conditions mean that, for the wavelength separation between the channels ~ 0.1 nm, a large peak power of the solitons is needed, ~ 1 W. While this effect is not practically important in terms of optical telecommunications, it is interesting for the study of the soliton dynamics. These bound states were studied in work [63] by means of the VA and direct simulations. The former was actually based on Eqs. (2.50) and (2.51), without the simplifying assumption (2.52). It was found that VA predictions compare quite well with direct numerical results. Studied were both the symmetric system, with equal PADs in the two channels (zero, anomalous, or normal), and asymmetric ones, with zero PAD in one channel and either anomalous or normal

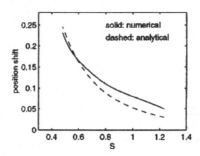

Figure 2.12: The analytically and numerically found position shift of the soliton induced by the complete collision in the two-channel DM model described by Eqs. (2.46) and (2.47) with $L_1 = 0.4$, $L_2 = 0.6$, $\beta_1 = -5/2$, $\beta_2 = 5/3$, $2c = 0.3$, peak powers of the colliding pulses being $P_u = P_v = 0.1$.

PAD in the other, or with opposite signs of the PADs in the two channels. In all the cases, it was found that stable bound states in which the solitons oscillate relative to each other exist indeed, provided that the energy of the solitons exceeds a certain minimum value $E_{\min}$, that depends on PADs and the inverse-group-velocity difference $2c$ between the channels. There is, however, a maximum value of $2c$, beyond which no bound state is possible. Typical examples of the dependences $E_{\max}(2c)$ are displayed in Fig. 2.13. Additionally, it was demonstrated in work [63] that, in the case when PAD in one channel is normal and so large that the DM soliton is unstable in it, the interaction with the soliton in the mate channel with anomalous PAD can produce a completely stable bound state.

2.6.3 Related problems

It is relevant to mention another two-channel model that, generally, also belongs to the class of the periodic heterogeneous nonlinear systems, although it does not involve DM. Instead, the group-velocity mismatch between the two XPM-coupled modes is subjected to periodic modulation, i.e., the model is based on the system of NLS equations (1.22), (1.23) with $\beta_u = \beta_v = \text{const} < 0$, $\gamma = \text{const} > 0$, $\sigma = 2$, and $c(z) = c_0 \sin(kz)$. A natural object in this model, considered in work [112], is a two-component symmetric soliton, with equal energies in both components. Without the modulation of $c(z)$, this compound soliton features an eigenmode of intrinsic excitation, in the form of mutual oscillations of centers of the two components, which was studied in detail [169, 87, 114]. The inclusion of the periodic modulation of the GVM between the components can give rise to resonant effects, if the modulation (spatial) frequency, $2\pi/k$, is commensurate with the eigenfrequency of the above-mentioned intrinsic mode. This possibility was explored in work [112], although only in the framework of the VA (without direct simulations of the coupled NLS equations). It was shown that the periodic modulation of $c(z)$ may split the compound soliton into two free single-component ones, the minimum GVM-modulation amplitude, $(c_0)_{\min}$,

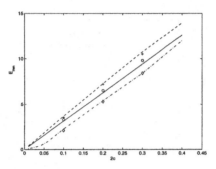

Figure 2.13: The minimum soliton's energy, necessary for the formation of bound states of solitons in the two-channel DM system, is shown as a function the inverse-group-velocity difference $2c$ for zero ($\beta_0^{(u)} = \beta_0^{(v)} = 0$), anomalous ($\beta_0^{(u)} = \beta_0^{(v)} = -0.1$), and normal ($\beta_0^{(u)} = \beta_0^{(v)} = 0.01$) path-average dispersion (PAD). The minimum energy predicted by the variational approximation for these three cases is shown, respectively, by solid, dashed-dotted, and dashed lines. Discrete symbols – circles, rhombuses, and crosses – represent values found from direct numerical simulations for the zero, anomalous, and normal PAD, respectively. The plots are aborted close to points where they abruptly shoot up (almost vertically); no bound state is possible for $2c > 0.40$ in the cases of the zero and normal PAD, and for $2c > 0.45$ with the anomalous PAD. In the asymmetric model, with $\beta_0^{(u)} \neq \beta_0^{(v)}$, the dependences $E_{\max}(2c)$ are quite similar.

necessary for the splitting, having deep minima (if considered as a function of k) at k corresponding to the fundamental and additional resonances with the intrinsic mode.

Interaction between DM solitons belonging to the same channel is also a problem of great interest (it is just "interaction", rather than collision, as the solitons keep a relatively large separation between themselves, the interaction being mediated by "tails" of each soliton overlapping with neighboring ones). In fact, it was found that this kind of the interaction gives rise to the most serious factor limiting the use of the DM, as, for relatively strong DM ($S \gtrsim 2.5$), interaction effects severely affect the maximum distance of the error-free transmission of data by soliton streams (see papers [171, 146] and references therein). The source of the problem is that, in the strong-DM regime, the solitons periodically spread out, which leads, through their overlapping and resulting formation of "ghost" pulses, to accumulation of mutual distortion induced by the FWM effect [116]. Semi-analytical consideration of the intra-channel interactions between DM solitons, based on a specially devised version of VA, was worked out and yielded quite accurate results (as compared to direct simulations), but the analysis is cumbersome. Details can be found in work [171].

Chapter 3

The split-step model

3.1 Introduction to the model

The most common numerical algorithms used for simulations of equations (and systems of equations) of the NLS type belong to the "split-step" type. It is based on splitting each step Δz of the numerical integration, $\Delta z = \Delta z_N + \Delta z_D$, so that only the nonlinear term(s) in the equation(s) are taken into regard at the first substep, and only the GVD and other (if any) linear term(s) are dealt with at the second substep. At the latter stage, the corresponding linear equation(s) are solved by means of the Fourier transform.

The *split-step model* (SSM), which was introduced in work [50] and further developed in paper [52], is formally similar to the split-step numerical algorithm, but the difference is that the propagation distances corresponding to Δz_N and Δz_D are not small, both being comparable to the soliton's period, see Eq. (1.16) (for this reason, they are denoted below as L_N and L_D, respectively, rather than as $\Delta z_{N,D}$). In other words, the SSM assumes periodic alternation of long segments of two different fiber species that are (in the first approximation) purely nonlinear and dispersive, respectively. Moreover, the nonlinear and dispersive components of the SSM are not necessarily fibers – the former may be SHG ($\chi^{(2)}$) modules [54], and the latter may be realized as short pieces of a fiber Bragg grating. Clearly, the SSM also belongs to the general class of the periodic heterogeneous nonlinear systems, and it is interesting to understand if SSM can support robust solitons.

SSM has something in common with previously studied fiber-optic schemes using the so-called comb-like dispersion profile, that assumes short segments of high-dispersion fiber inserted in a low-dispersion bulk fiber [161]. However, there is strong difference of SSM from the "comb" schemes: in the latter case, a large number ($\simeq 8$) of strong-dispersion segments are inserted (non-uniformly) per dispersion length, with the objective to emulate a continuous exponentially decreasing dispersion profile, adjusted to the gradual decay of the soliton's energy due to the fiber loss (so that the soliton does not feel the system's heterogeneous structure, and its temporal width remains nearly constant). On the contrary, the SSM typically assumes one dispersive and one nonlinear sections per dispersion length, and the transmission regime is completely different from that in the comb systems.

SSM offers a potential for applications to optical telecommunications. On the one hand, periodically inserting short strongly nonlinear elements can help to upgrade a linear fiber-optic link, making it possible to transmit solitons through it. On the other hand, periodic insertion of strong dispersive elements can be useful to improve pulse transmission in links using dispersion-shifted fibers (ones with a small value of the GVD coefficient), where the nonlinearity may be too strong versus the dispersion.

3.2   Solitons in the split-step model

3.2.1   Formulation of the model

The dispersive segment of the SSM is described by the linear version of the NLS equation (1.5), $iu_z + (1/2)u_{\tau\tau} = -i\alpha_D u$, where the GVD coefficient is normalized to be $\beta = -1$ (in the case of normal dispersion in the dispersive segments, $\beta > 0$, the SSM supports no solitons), and α_D is the loss constant of the dispersive fiber (in physical units, typical values of β and α_D in standard telecommunication fibers are -20 ps^2/km, and 0.2 dB/km, respectively). The substitution of $u(z, \tau) \equiv v(z, \tau) \exp(-\alpha_D z)$ leads to the lossless equation for the dispersive segment,

$$iv_z + \frac{1}{2}v_{\tau\tau} = 0, \tag{3.1}$$

that can be solved by the Fourier transform in τ.

In the nonlinear segment, one is dealing with the dispersionless version of equation (1.5),

$$iu_z + |u|^2 u = -i\alpha_N u, \tag{3.2}$$

where the nonlinearity coefficient is normalized to be $\gamma = 1$ (its typical physical value in optical fibers is 2 (W·km)$^{-1}$), and α_N is the respective loss parameter. An obvious solution to Eq. (3.2) is

$$u(z, \tau) = u(0, \tau) \exp\left(-\alpha_N z + i\frac{|u(0, \tau)|^2}{2\alpha_N}\left[1 - \exp(-2\alpha_N z)\right]\right). \tag{3.3}$$

In the lossless limit, $\alpha_N \to 0$, it takes the form

$$u(z, \tau) = u(0, \tau) \exp\left(i|u(0, \tau)|^2 z\right). \tag{3.4}$$

It is assumed, as usual, that the losses are compensated by linear amplifiers which act on the wave field so that

$$u(\tau) \to u(\tau) \cdot e^G, \tag{3.5}$$

where G is the gain (in applications, the gain is usually measured in dB (deciBells), which will be 8.69G). The value of the gain is selected so that to provide for the balance with the total loss,

$$G = L_N \alpha_N + L_D \alpha_D, \tag{3.6}$$

where L_D and L_N are lengths of the periodically alternating dispersive and nonlinear segments. In fact, the model is equivalent to its lossless version ($\alpha_D = \alpha_N = G = 0$) with a renormalized value of L_N. Indeed, comparing Eqs. (3.3) and (3.4), and taking Eqs. (3.5) and (3.6) into regard, it is easy to see that the model including the losses and gain is tantamount to the lossless one with L_N replaced by

$$L_N^{(\text{eff})} = (2\alpha_N)^{-1}\left[\left(1 - e^{-2\alpha_N z_0}\right) + e^{2G - 2\alpha_N z_0}\left(1 - e^{-2\alpha_N(L_N - z_0)}\right)\right], \quad (3.7)$$

where z_0 is the distance from the beginning of the nonlinear segment to the point at which the amplifier is installed. For this reason, in what follows below only the lossless model is considered, making no distinction between $L_N^{(\text{eff})}$ and L_N, nor between u and v, see Eq. (3.1). Dynamical invariants of the lossless SSM are the same energy and momentum as defined above in Eqs. (1.9) and (1.10). In the absence of losses, Eqs. (3.1) and (3.2) are invariant with respect to the transformations, respectively

$$\tau \quad \rightarrow \quad \tau/\Lambda_D,\, z \rightarrow z/\Lambda_D^2,$$
$$u \quad \rightarrow \quad \Lambda_N u,\, z \rightarrow z/\Lambda_N^2 \qquad (3.8)$$

with arbitrary rescaling factors Λ_D and Λ_N. This transformation may be used to set, for instance, $L_N = L_D = 1/2$, so that the full size of the system's cell is $L = 1$.

The exact definition of the cell is an interval between midpoints of two neighboring nonlinear segments, with a dispersive segment inserted between them. A full transformation (*map*) of the pulse passing the system's cell can be represented as a superposition of two transformations (3.4) corresponding to the nonlinear half-segments at edges of the cell, and the linear transform between them, corresponding to the dispersive segment in the middle of the cell. Numerical simulations of the pulse evolution in the SSM are performed by many iterations of the map, with a fixed cell's size in the case of the regular system (or with the values of the size picked randomly from a finite interval for a *random SSM*, see below).

Note that averaged (in z) version of both the regular and random SSM systems amounts to the usual NLS equation,

$$2iu_z + \frac{1}{2}u_{\tau\tau} + |u|^2 u = 0 \qquad (3.9)$$

For this reason, it is natural to start the simulations with an initial pulse which would be a fundamental soliton of the average equation (3.9),

$$u_0(\tau) = \eta \operatorname{sech}(\eta\tau), \qquad (3.10)$$

with an arbitrary amplitude η. Besides that, in order to understand the operation of the system in the general case, initial pulses with an arbitrary relation between the amplitude and width will also be considered,

$$u_0(\tau) = \eta \operatorname{sech}\left(\frac{\eta\tau}{W}\right), \qquad (3.11)$$

where W is a relative width parameter. Note that, in the case of the ordinary NLS equation, an asymptotic (for $z \rightarrow \infty$) form of the solution generated by the generic initial pulse (3.11) can be found in an exact form for any value of W [154]. With the GVD and nonlinearity coefficients fixed as in Eqs. (3.2) and (3.1), and with the above normalization, $L_N = L_D = 1/2$, free remaining parameters of the model are the amplitude η and relative width W of the initial pulse (3.11).

It should also be noted that, while the dispersive equation (3.1) is invariant with respect to the Galilean transformation (1.6), the nonlinear equation (3.2) is not formally Galilean-invariant; however, it is obvious that the system as a whole is invariant with respect to a modified boost transformation,

$$u(z,\tau) \mapsto u(z, \tau - c\tilde{z}) \exp\left(-c^2\tilde{z}/2 + ic\tau\right), \tag{3.12}$$

where $\tilde{z}$ is the distance passed only in the dispersive segments. As well as in the case of the ordinary NLS equation, the effective Galilean invariance of the SSM is related to the conservation of the momentum (1.10).

3.2.2 Variational approximation

To apply the variational approximation (VA) to SSM solitons, the usual ansatz (2.6) can be used, and, in the nonlinear segment, Eqs. (2.10) – (2.13) yield the following evolution equations for the width and chirp:

$$\frac{db}{dz} = -\frac{2}{\pi^2}\frac{E}{a^3}, \ \frac{da}{dz} = 0, \ \frac{dE}{dz} = 0 \tag{3.13}$$

(here, as above, the definition of the energy is $E = A^2 a$). In the dispersive segment, the variational equations amount to

$$\frac{d^2a}{dz^2} = \frac{4}{\pi^2}\frac{1}{a^3}, \ b = \frac{1}{2a}\frac{da}{dz}. \tag{3.14}$$

As E is a constant, a solution to Eqs. (3.13) is trivial,

$$a = \text{const} \equiv a_{\max}, \ b(z) = -\frac{2E}{\pi^2 a_{\max}^3}(z - z_N), \tag{3.15}$$

where z_N is an arbitrary constant. A general solution to Eqs. (3.14) is simple too,

$$a(z) = \frac{\sqrt{\left(\pi a_{\min}^2\right)^2 + 4\left(z - z_D\right)^2}}{\pi a_{\min}}, \tag{3.16}$$

$$b(z) = \frac{2\left(z - z_D\right)}{\left(\pi a_{\min}^2\right)^2 + 4\left(z - z_D\right)^2}, \tag{3.17}$$

where z_D and $a_{\min}$ are other arbitrary constants.

In terms of the VA, steady transmission of the soliton implies that its amplitude, width, and chirp return to their original values after passing a full cell of the system. It immediately follows from Eqs. (3.15) and (3.16) that, if z_D is chosen to coincide with the midpoint of the dispersive segment, the width $a(z)$, which takes the value $a_{\min}$ at this midpoint, automatically returns to the same value at the midpoint of the next dispersive segment, $a_{\min}$ being the smallest value which the width of the pulse attains in the course of its periodic vibrations.

According to Eq. (3.17), the soliton's chirp vanishes at the midpoint of the dispersive segment. A condition which guarantees that it vanishes at the midpoint of the next dispersive segment can be easily obtained: the net change of $b(z)$ in the nonlinear segment, which is

$$(\Delta b)_N = -\frac{2E}{\pi^2 a_{\max}^3} L_N \qquad (3.18)$$

according to Eq. (3.15), must exactly compensate the difference of the values of the chirp at edges of the dispersive segment, which is, according to Eq. (3.17),

$$(\Delta b)_D = \frac{2L_D}{\left(\pi a_{\min}^2\right)^2 + L_D^2} \qquad (3.19)$$

(to make the meaning of the expressions clearer, it is not assumed here that $L_D = L_N$). Note that, due to the continuity of $a(z)$, the value $a_{\max}$ which appears in Eq. (3.18) is one attained at the edge of the dispersion segment, i.e., it is given by Eq. (3.16) with $z - z_D = L_D/2$,

$$a_{\max} = \frac{\sqrt{\left(\pi a_{\min}^2\right)^2 + L_D^2}}{\pi a_{\min}} . \qquad (3.20)$$

In fact, $a_{\max}$ is indeed the maximum width attained in the process of the steady propagation of the SSM soliton.

Finally, the substitution of Eqs. (3.18), (3.20), and (3.19) into the balance condition for $b(z)$, $(\Delta b)_N + (\Delta b)_D = 0$, yields a basic result:

$$\pi^2 \left(\frac{L_N}{L_D} E\right)^2 a_{\min}^6 = \pi^2 a_{\min}^4 + L_D^2 . \qquad (3.21)$$

This is a *constitutive equation* for the SSM solitons, which (as predicted by the VA) determines its minimum width, $w_{\min}$, as a function of the energy E. This equation can also be written in terms of the maximum width,

$$\pi \left(\frac{L_N}{L_D} E\right) a_{\max}^3 = \pi a_{\max}^2 + (L_N E)^2 . \qquad (3.22)$$

Considering E as a function of $a_{\min}$ or $a_{\max}$, it is easy to see that Eqs. (3.21) and (3.22) yield exactly one physical (real) value of $a_{\min}$ and exactly one value of $a_{\max}$ for any $E > 0$.

In the limit of $L_D, L_N \to 0$, the SSM reduces to the average NLS equation (3.9), with the extra coefficient L_D/L_N in front of the term $(1/2) u_{\tau\tau}$. In this limit, the width $a(z)$ becomes constant, $a = a_{\min} = a_{\max}$. On the other hand, the exact fundamental-soliton solution (1.13) of the thus defined NLS equation has

$$|u(\tau)| = \sqrt{\frac{L_D}{L_N}} \frac{1}{a_0} \operatorname{sech}\left(\frac{\tau}{a_0}\right) \qquad (3.23)$$

with an arbitrary width a_0, the soliton's energy being $E = (L_D/L_N) a_0^{-1}$. Inserting this in the limiting forms of Eqs. (3.21) and (3.22) corresponding to $L_D, L_N \to 0$, one concludes that they are satisfied automatically, i.e., the VA correctly reproduces the exact result for the fundamental soliton in the NLS limit. The same limit corresponds to $E \to 0$ at finite L_D and L_N, as in this case the soliton becomes very broad, with the dispersion length $Z_D \sim a^2 \gg L_D, L_N$, hence it must be asymptotically equivalent to the ordinary NLS soliton.

It is also relevant to note that, as it follows from Eq. (3.21), the minimum width a_{min} may take any value from 0 to ∞ when E is varied from ∞ to 0. However, Eq. (3.22) shows that the maximum width a_{max} diverges in both limits, $E \to 0$ and $E \to \infty$, which suggests that there is a finite smallest value that a_{max} may assume. Indeed, analysis of Eq. (3.22) demonstrates that this value is $(a_{max})_{min} = \sqrt{(2/\pi) L_D}$, and it is attained at $E = \sqrt{2\pi L_D} L_N^{-1}$. In this case, the minimum width is $(1/\sqrt{2}) (a_{max})_{min}$.

3.2.3   Comparison with numerical results

Direct simulations of the SSM were performed in works [50, 52]. The simulations started with the initial wave form (3.10) that would generate a fundamental soliton in the averaged NLS equation corresponding to the SSM. In the case when the soliton period in the averaged equation (defined as per Eq. (1.16)) is comparable to the cell's size L, a soliton readily self-traps in the SSM, with an extremely small radiation loss and very little change of the shape against the initial form, see an example in Fig. 3.1.

If the opposite case, with L much larger than the soliton period (with the latter defined as per the average NLS equation, see Eq. (1.16)), the adjustment of the soliton and radiation loss accompanying its relaxation to the eventual shape are quite conspicuous, as shown in Fig. 3.2. In this case, the established soliton features intrinsic *chirp* (i.e., the wave field is complex), as shown in Fig. 3.2(d). Nevertheless, the amplitude distribution in the soliton, $|u(\tau)|$, is still well fitted by the usual sech ansatz, see Fig. 3.2(c). In fact, the proximity of the pulse to the classical shape of the NLS soliton may be characterized by its *area*,

$$A \equiv \int_{-\infty}^{+\infty} |u(\tau)| \, d\tau \tag{3.24}$$

(unlike the energy (1.9), the area is not a dynamical invariant of the NLS equation). For any soliton of the average NLS equation (3.9), $A = \pi$ (note that the area does not depend on the soliton's amplitude). For the established soliton displayed in Fig. 3.2, the area is 3.25, i.e., quite close to π. If L is very large in comparison with the soliton period in the average NLS equation, the initial pulse completely decays into radiation, as will be explained in more detail below.

To compare the analytical results outlined above with direct simulations in a systematic way, it is more convenient to cast the constitutive equation (3.21) in a different form, which determines the maximum amplitude of the SSM soliton, $A_{max} = \sqrt{E/a_{min}}$ (recall that $E = A^2 a$; obviously, the largest amplitude is achieved at a point

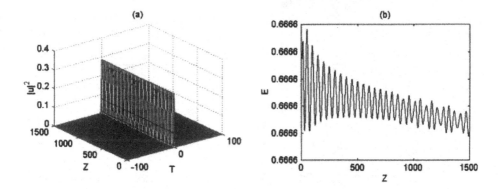

Figure 3.1: (a) Numerically simulated evolution of a soliton in the split-step model with the cell's size $L = 1$, starting with the initial configuration $u_0 = \mathrm{sech}\,\tau$ (which generates a soliton in the average NLS equation (3.9) with the soliton period $\pi/2$). (b) Evolution of the soliton's energy in the course of its adjustment to the established shape. The total propagation distance shown in the figure corresponds to 1500 system's cells.

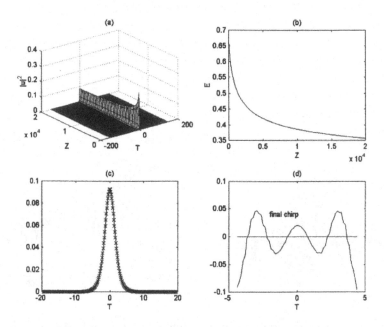

Figure 3.2: The same as in the previous figure, but with the SSM period ten times as large, $L = 10$. Panels (c) and (d) additionally show the fit of the eventual shape of the soliton to the usual sech form, and the distribution of the local *chirp*, $d^2\phi/d\tau^2$, in the established soliton ($\phi(\tau)$ is the phase of the soliton's complex wave field).

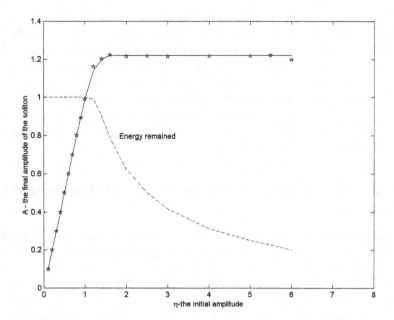

Figure 3.3: Continuous curve shows the largest amplitude $A_{\max}$ of the soliton in the split-step model, as predicted by the variational approximation, see Eq. (3.25), vs. the amplitude η of the initial pulse (3.10), in the case of $L_D = L_N = 1/2$. The dashed curve shows the share of the initial energy which remains trapped in the established soliton, after the completion of its relaxation, as found from numerical data (this residual energy was used to generate the continuous curve). Stars show a set of values of the amplitude of the established soliton at midpoints of dispersive segments, as found from direct simulations.

where the soliton's width is smallest, $a(z) = a_{\min}$):

$$L_D^2 \left(A_{\max}^4 \right)^3 + \pi^2 E^4 A_{\max}^4 = \pi^2 \left(\frac{L_N}{L_D} \right)^2 E^8. \qquad (3.25)$$

This prediction for the amplitude was compared to results of direct simulations starting with the exact soliton (3.10) of the average NLS counterpart of the SSM, Eq. (3.9). The comparison is presented in Fig. 3.3 (for the case of $L_D = L_N = 1/2$, which is tantamount to the general case, as explained above). It is necessary to take into regard that, because of the radiative loss suffered by the soliton in the course of its adjustment to the established regime, the energy of the finally observed SSM soliton may be considerably smaller than the energy $E_{\text{in}} \equiv \eta$ of the initial pulse (3.10), as is seen, for instance, in Fig. 3.2(b). Therefore, Fig. 3.3 was "phenomenologically" improved, by substituting E in it by the value of energy found from numerical data for the established soliton (the share of the initial energy which remains trapped in the soliton is also shown in Fig. 3.3).

A noteworthy feature of the dependence of the SSM-soliton's amplitude vs. the amplitude of the initial pulse is *saturation* obvious in Fig. 3.3. If one launches the pulse with a large amplitude, it quickly sheds off a considerable part of its energy in the form of radiation waves, and relaxes to the eventual form. Note that the characteristics shown in Fig. 3.3, are universal, as they do not depend on any remaining free parameter.

A feature which is not predicted by VA is *termination* of the characteristics: the curve in Fig. 3.3 is not aborted arbitrarily, but ends at a point beyond which no stable SSM soliton is produced by simulations. It was observed that, past the termination point, the solitons disappear abruptly.

3.2.4 Diagram of states for solitons and breathers in the split-step system

General results characterizing the dynamics of solitons and quasi-solitons in the SSM can be collected from systematic simulations starting with a pulse (3.11), which admits an arbitrary relation between the amplitude and width (controlled by the parameter W), rather than locking them to the form of the average soliton (3.10). It is well known that, in the case of the ordinary NLS equation, the evolution problem for the initial condition (3.11) has an exact analytical solution, in terms of the inverse scattering transform [154]. The latter solution demonstrates that the configuration (3.11) generates no soliton if $W < 1/2$; in the interval $1/2 < W < 3/2$, a fundamental soliton is generated, together with some amount of radiation; and higher-order n-solitons are produced in intervals $n - 1/2 < W < n + 1/2$ (which is also accompanied by emission of radiation, unless W is an integer). The higher-order solitons, unlike the fundamental one, look like breathers, demonstrating persistent internal vibrations (see the expression for the 2-soliton given in Eq. (1.15).

Results of systematic simulations performed in the SSM with the initial condition (3.11) are summarized in the diagram displayed in Fig. 3.4, which clearly shows similarities and differences between the NLS and SSM models. Regions generating qualitatively different states, *viz.*, a fundamental soliton, a breather, and separating pulses ("splitting"), are identified in the diagram. The white area is one where the initial configuration completely decays into dispersive radiation, without generating any persistent localized state. Delineating all the borders in the diagram in a very accurate way would demand an extremely large number of very long simulations, therefore some borders have a rather approximate form.

The lower horizontal border which marks the threshold of the fundamental-soliton formation is virtually the same as the above-mentioned one, $W = 1/2$, in the NLS equation. However, a drastic difference between the SSM and NLS equation is that, at large η, no soliton is generated. In particular, the plot shown in Fig. 3.3 terminates at a point which corresponds to the intersection between the right border of the soliton region in Fig. 3.4 and the line $W = 1$ (Fig. 3.3 was generated for this value).

At small η, the breather-formation border is almost the same as the above-mentioned one in the NLS equation, i.e., $W = 3/2$. At larger η, the difference of SSM from the NLS equation manifests itself in the uplift of the border to $W \approx 2$. Moreover, the fundamental-soliton region protrudes farther upward in the interval $3 < \eta < 4$,

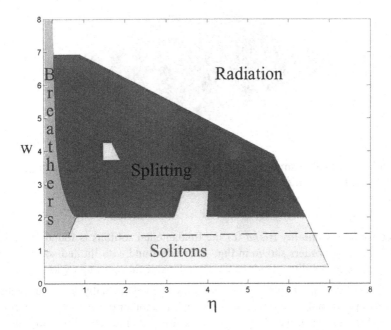

Figure 3.4: The diagram showing different outcomes of the evolution of the initial pulse (3.11) in the split-step model with $L_D = L_N = 1/2$. In white area the initial pulse completely decays into radiation. The dashed horizontal line, $W = 3/2$, is the exact breather-generation threshold in the corresponding averaged NLS equation (3.9).

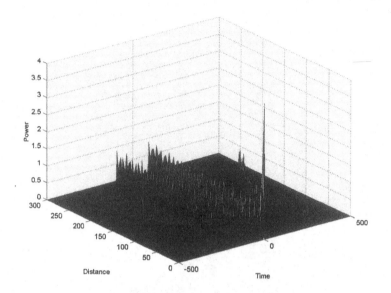

Figure 3.5: A typical example of splitting of the initial pulse (3.11), with $\eta = 2$ and $W = 3$, into a set of stable separating breathers.

and a conspicuous *stability island* for the fundamental solitons is found around $\eta = 1.5, W = 4$. The borders shown in Fig. 3.4 were found with limited accuracy; quite plausibly, smaller stability islands can be found inside the splitting region.

Splitting of the initial pulse into several moving ones (actually, they are breathers), which takes place in the large region in Fig. 3.4, is another drastic difference from the NLS equation, in which the chirpless initial pulse (3.11) never splits. Varying η and W, one may observe splitting into up to seven moving breathers (if the number of the splinters is odd, the central one remains quiescent). A typical example of the splitting into four fragments is displayed in Fig. 3.5. The splitting of a moderately broad initial pulse may be prevented if chirp of a proper sign and magnitude is added to it [50]. In this case, the pulse eventually transforms itself into a single soliton, which keeps almost all the initial energy and has almost no chirp. This property is another essential difference of SSM from the NLS equation.

In the narrow region where a single stable breather is generated, it features irregular long-period oscillations, as shown in Fig. 3.6 (the length along the z-axis in this figure is given in units of the SSM period L). Inspection of this example shows that the pulse periodically assumes a double-peaked shape; multi-peaked breathers were found at larger values of W. At small values of η, the breather becomes unstable against splitting into separating pulses, which are breathers too (not shown here; examples can be found in [52]). In the latter case, the splitting takes place after a long quasi-stable evolution stage, and some spontaneous symmetry breaking is observed.

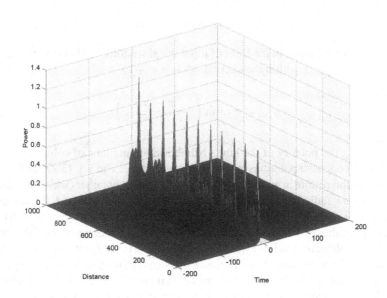

Figure 3.6: An example of formation of a stable quasi-periodic breather from the initial pulse (3.11) in the case of $\eta = 0.4$ and $W = 3$.

It was mentioned above that the initial pulse corresponding to the soliton of the average-NLS counterpart of the SSM, in the form (3.10) with $\eta \sim 1$, completely decays into radiation in the SSM with very large L. In terms of Fig. 3.4, which pertains to $L \equiv 1$, the latter is tantamount to taking very large η. Indeed, the figure shows that such initial pulses suffer complete decay, which is another salient difference from the NLS equation (where the same initial pulse would result in formation of a soliton of a very high order).

The diagram in Fig. 3.4 does not include small stability islands for solitons and breathers (except for the aforementioned one, found around the spot with $\eta = 1.5, W = 4$), which definitely exist inside the "radiation" region. Actually, a complex (plausibly, fractal) system of stability islands was found in Ref. [50] in the case of $W = \eta = 1$, by increasing the value of L in steps of $\Delta L = 1$. As a result, it was found that the soliton remains continuously stable up to $L = 14$; then, stability islands were found around the following values of L:

$$L = 18, \; L = 20, \; L = 22, \; L = 24, \; L = 26, \; L = 36, \; L = 51, \; L = 59. \qquad (3.26)$$

In the islands at large L, the stable solitons have a small amplitude, and are broad. No stability regions were found for $L \geq 60$.

In connection to the latter finding, it is relevant to mention that a sophisticated system of alternating windows was earlier discovered in a very different nonlinear model based on the Goldstone equation (also called ϕ^4 equation) for a real function $\phi(x, t)$,

$$\phi_{tt} - \phi_{xx} - \phi + \phi^3 = 0,\qquad(3.27)$$

which has exact solutions in the form of topological solitons (kinks) that may move at an arbitrary velocity c, within the interval $-1 < c < +1$,

$$\phi_{\text{kink}} = \sigma \tanh\left(\frac{x - ct}{\sqrt{2(1 - c^2)}}\right),\qquad(3.28)$$

$\sigma = \pm 1$ being the polarity of the kink. Numerical simulations of collisions between the kinks with opposite polarities and opposite velocities $\pm c$ in Eq. (3.27) had revealed that the collision results in annihilation of the kinks into a breather (which is subject to subsequent slow radiative decay) if c is very small. On the other hand, the collision is quasi-elastic (i.e., the kinks pass each other with almost no loss) if c is sufficiently close to 1. Between these two cases, a system of alternating windows of annihilation and quasi-elastic collisions was found [36]. A semi-qualitative explanation to these findings was based on the collision-induced exchange between the original kinetic energy of the kinks, and the energy absorbed by the internal oscillatory degree of freedom, which the kink is known to have in this model (internal oscillations are excited in both kinks as a result of the collision). A similar fine system of alternating windows was found, and a similar explanation to it was proposed, for kink-antikink collisions in the so-called double sine-Gordon equation [37]. It may happen that the SSM soliton also has an internal degree of freedom (note that the ordinary NLS soliton does not have any intrinsic mode). However, the latter issue has not been explored.

3.3   Random split-step system

As was mentioned above in connection to the random DM systems, the study of heterogeneous nonlinear models in which cells alternate not periodically but randomly (i.e., the length of the cell is picked randomly from some interval, such as (2.32)) is an issue of significant interest to applications, and in its own right as well. The random version of the SSM was considered in work [52]. To pose the random-SSM model, it was assumed that the lengths L_N and L_D of the nonlinear and dispersive segments are not fixed as above (for instance, as $L_N = L_D = 1/2$), but are taken at random from an interval $L_{\min} < L_N = L_D \equiv L/2 < L_{\max}$ (note that it is assumed that the lengths L_N and L_D remain mutually locked in each cell; if, instead, they are chosen at random and independently from each other, no stable soliton can be found in the system).

Numerous runs of simulations, performed for the random model at various values of parameters, have yielded a simple conclusion: if the ratio $L_{\max}/L_{\min}$ is not very large (for instance, if $L_{\max}/L_{\min} = 5$), stable SSM solitons persist in the random system, and, on the average, they seem almost identical to the solitons in the regular (strictly periodic) SSM model, with L simply equal to the median value, $\overline{L} = (1/2)(L_{\max} + L_{\min})$, of the random distribution. As a typical example, Fig. 3.7 displays the evolution of the initial pulse (3.10) with $\eta = 1$ in the random SSM with $L_{\min} = 1$,

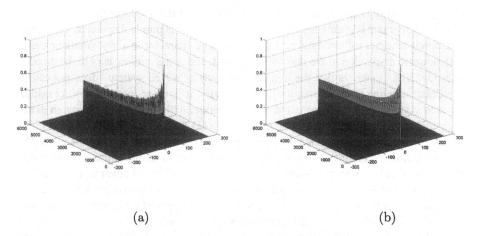

(a)   (b)

Figure 3.7: (a) A typical example of the formation of a soliton in the random split-step model, with $L_{\min} = 1$ and $L_{\max} = 5$, from the initial pulse (3.10) with $\eta = 1$. (b) For comparison, the same is shown in the regular system with $L = (1/2)\,(L_{\max} + L_{\min}) \equiv 3$.

$L_{\mathrm{max}} = 5$, and, for comparison, the evolution of the same initial pulse in the periodic SSM with $L = \overline{L} \equiv 3$.

Thus, SSM solitons are stable against the disorder (in fact, they seem more robust, in this sense, then the DM solitons). Moreover, Fig. 3.7 (and many more numerical results) clearly shows that, as well as in the regular system, the soliton in the random SSM is an effective *attractor*, in the sense that initial pulses whose parameters, such as the amplitude or initial chirp, are different from those of an SSM soliton, quickly relax to it (in particular, suppression of the initial chirp can be observed). Attractors are typical to dissipative systems; nevertheless, in conservative nonlinear-wave systems they are possible too, due to the effective dissipation through radiation losses.

Other properties of the solitons in the disordered SSM mode and its regular counterpart were also shown to be very similar – such as the stability limits (see Fig. 3.4), formation of breathers, etc.

3.4   A combined split-step – dispersion-management system: dynamics of single and paired pulses

The above consideration was dealing with a limit case of the SSM, composed of cells in which one segment is purely nonlinear, and the other one is purely dispersive. A natural question is whether soliton dynamics keeps the same character in a more realistic case, with nonzero dispersion in the nonlinear segment, and some nonlinearity in the dispersive one. The answer is that the transition to such a "mixed system" does not cause any drastic change in the dynamics [53]. Actually, it is more interesting to consider a mixed system which is constructed as a DM model with the map including an extra segment, with strong nonlinearity and negligible GVD. As was shown in work [53], such a *three-step* system, which may be regarded as a combination of SSM and DM, produces quite interesting and promising (for applications) results, although robust RZ pulses found in it are not true solitons. In particular, an essential advantage offered by the combined SSM-DM system is prevention of developing the overlap between adjacent pulses, see below (the latter problem is known in optical telecommunications as *inter-symbol interference*, or ISI). It is relevant to mention that three-step versions of the DM provide some other assets, such as better suppression of effects of inter-channel collision between RZ pulses (not necessarily solitons) in the WDM setting [17].

3.4.1   The model

The combined SSM-DM system is based on Eq. (1.48) with the map that involves the piecewise-constant modulation of both β and γ :

$$\{\beta, \gamma\} = \begin{cases} \{\beta_1 = 0, \gamma_1\} & \text{if } 0 < z < L_1, \\ \{\beta_2, \gamma_0\} & \text{if } L_1 < z < L_1 + L_2, \\ \{\beta_3 = -\beta_2, \gamma_0\} & \text{if } L_1 + L_2 < z < L_1 + L_2 + L_3. \end{cases} \tag{3.29}$$

The three-step cell (3.29) repeats itself with the period $L \equiv L_1 + L_2 + L_3$. Here, γ_0 and γ_1 are, respectively, the nonlinearity of the system fiber, and of the additional strongly nonlinear segment.

The latter (additional nonlinear) element may be realized in several different ways. One possibility is to use a long segment of a dispersion-shifted fiber, with a usual value of the nonlinear coefficient and very weak dispersion. Another realization may be based on a short (less than a meter) piece of an Erbium-doped fiber, which, if properly designed and pumped, may feature the nonlinearity coefficient larger by a factor of $\simeq 5 \cdot 10^5$ than in the regular fiber. Also quite short may be a segment of a PCF (photonic-crystal fiber), which can provide for very strong nonlinearity with diverse arrangements of the fiber's microstructure, see paper [124] and references therein. Moreover, the nonlinear element must not necessarily be a piece of a fiber; instead, it may be a compact module, based on an SHG crystal, in which strong effective $\chi^{(3)}$ nonlinearity is induced by $\chi^{(2)}$ interactions through the cascading mechanism, while the module's GVD may be completely neglected in view of its small size (a detailed description of the model with SHG modules is given in the following chapter, which deals with the nonlinearity management). Loss and gain terms are not included in Eq. (3.29), following the usual assumption of the local compensation of the fiber loss by lumped amplifiers.

The simulations start with the unchirped Gaussian pulse at $z = 0$,

$$u_0(t) = \sqrt{P_0} \exp\left(-\frac{t^2}{T^2}\right), \qquad (3.30)$$

with the peak power P_0 and width T. Following the analogy with the definition of the DM strength (2.25), it is convenient to define a dimensionless *nonlinearity strength* of the extra segment,

$$NS \equiv \gamma_1 P_0 L_1 \qquad (3.31)$$

(actually, it is the the nonlinear phase shift at the center of the pulse passing the nonlinear segment).

3.4.2 Transmission of an isolated pulse

The aim of the consideration of the combined SSM-DM system is not to construct a true soliton solution, but rather to find propagation regimes for robust RZ pulses that may outperform the operation mode with usual DM solitons. Systematic simulations of the model with the initial condition (3.30) demonstrate that, in a broad range of parameter values, the propagation leads to *self-compression* of the pulse, with simultaneous generation of side-lobes attached to it in the temporal domain, a typical example of which is displayed in Fig. 3.8 (as the pulse performs nearly periodic shape oscillations, the figure shows it at a point where it is narrowest). The self-compression of the pulse is an effect of the additional nonlinearity added to the system. It can be checked that the peak power of the pulse in the case shown in Fig. 3.8 is too small for the formation of a soliton, but large enough to make the nonlinearity effects significant. Without the nonlinear segment inserted into the DM map, no systematic reduction of the width is observed as a result of the transmission.

The example shown in Fig. 3.8, as well as many others, suggest that, up to some value of the transmission length, the overall quality of the pulse improves, as its width

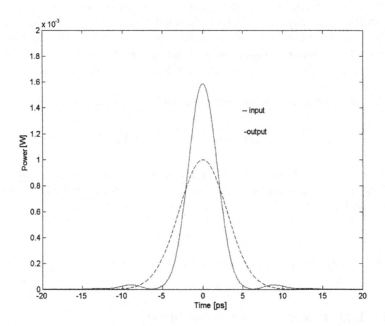

Figure 3.8: A typical example of comparison of the input Gaussian pulse and the output one (in the present case, it was produced by the propagation through 30 maps of the combined SSM-DM system). The side-lobes of the output pulse contain, in this case, only1.6% of the total energy.

gets reduced, roughly, by a factor of 2; after that, although the self-compression of the central body of the pulse continues, its overall quality starts to deteriorate due to the growth of the side peaks. A trade-off between these two trends defines the maximum acceptable (*optimal*) transmission length z_{opt}. Analysis of the numerical data reveals that z_{opt} virtually does not depend on the DM strength (2.25), when the latter takes values in a very broad interval,

$$1.5 < S < 11 \tag{3.32}$$

(outside this interval, the results are much worse) [53]. However, the optimal propagation length is quite sensitive to the nonlinearity strength (3.31) of the additional segment: the best performance is observed at $NS \approx 0.05$, and at $NS \gtrsim 0.10$ the system becomes inferior to its usual DM counterpart.

For further understanding of the dynamics of the pulses in the SSM-DM model, it is necessary to know how much they spread out in the course of the transmission. To this end, Fig. 3.9 displays plots showing the evolution of the pulse's width within one cell in the present model, and in its DM counterpart (the one without the extra nonlinear segment). For this figure, the integral definition of the squared temporal half-width is adopted,

$$T_{int}^2 \equiv \frac{\int_{-\infty}^{+\infty} t^2 |u(t)|^2 dt}{\int_{-\infty}^{+\infty} |u(t)|^2 dt}. \tag{3.33}$$

As is evident from Fig. 3.9, the same initial configuration produces a pulse that, on average, is definitely narrower in the present system than its counterpart in the ordinary DM model.

In work [53] it was also shown that the combined system provides for more efficient suppression of the temporal jitter of the RZ pulse (induced by random optical noise) than the DM model per se. This beneficial effect may be due the fact that, inside the additional nonlinear segment, the phase shift acquired by the pulse is much larger than that of the small-amplitude noise components, which makes the interaction between the pulse and the noise effectively incoherent, i.e., weak.

3.4.3 Transmission of pulse pairs

As mentioned above, a key problem hampering the use of strong-DM schemes in the soliton regime is the intra-channel interaction between solitons. To understand if the SSM-DM model alleviates this difficulty, it is necessary to simulate the co-propagation of a pair of two pulses, created with a temporal delay Δt. The objective is to find the minimum value of Δt that provides stable coexistence of the pulses (in particular, without conspicuous shifts of their centers due to the interaction). It was found [53] that, in the interval (3.32), the SSM-DM system gives rise to the minimum separation

$$(\Delta t)_{min} = 1.57 \, T_{FWHM}, \tag{3.34}$$

where if $T_{FWHM} = 1.18 \, T$ is the standard width of the Gaussian pulse (3.30), within the propagation distance where it is possible to maintain the acceptable quality of the

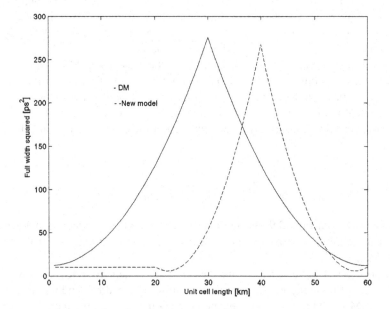

Figure 3.9: Comparison of the evolution of the squared half-widths of the pulse (generated by the same input) within one map (system's cell), in the combined SSM-DM system and its DM counterpart, that does not include the extra nonlinear segment. The integral definition (3.33) of the squared half-width is adopted here. The two plots are juxtaposed so that the borders between the anomalous- and normal-GVD segments, where the pulse's width attains its maximum in the ordinary DM system, coincide. In the plot corresponding to the SSM-DM model, the width keeps a small constant value inside the additional nonlinear segment ($0 < z < 20$).

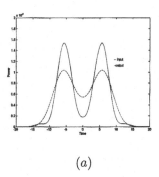
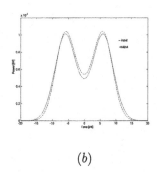

(a) (b)

Figure 3.10: The comparison between the input and output two-pulse configurations in the combined SSM-DM system (a) and its ordinary DM counterpart (that does not include the extra nonlinear segment) (b). The output is generated by the transmission of the pair through 30 system's cells.

single-pulse transmission, see above. If Δt exceeds $(\Delta t)_{\min}$, the co-propagating pulses feature virtually no interaction at all; in the opposite case, $\Delta t < (\Delta t)_{\min}$, the pulses merge into a single one.

The small value of $(\Delta t)_{\min}$ is quite promising for the applications, making it possible to realize a high bit rate (per channel) in the fiber-optic telecommunication link. For instance, for the pulse width $T_{\mathrm{FWHM}} = 7.08$ ps, that was actually used in the above examples, Eq. (3.34) yields $(\Delta t)_{\min} = 1.57\, T_{\mathrm{FWHM}} = 11.12$ ps, which implies the maximum bit rate as high as 89 Gb/s per channel.

In fact, the present system not only prevents the merger of the pulse pair with $\Delta t > (\Delta t)_{\min}$, but also improves the quality of the double-pulse configuration, showing a trend to clear the space between them, which means suppression of the above-mentioned ISI effect. The latter property is illustrated by typical examples in Fig. 3.10, through comparison between the input and output shapes of the two-pulse configurations, in the case of $\Delta t = 1.69\, T_{\mathrm{FWHM}}$, which is close to the minimum necessary separation, as per Eq. (3.34). In the figure, the comparison is given, in parallel, for the full model and its DM counterpart, which makes the effect of the ISI suppression obvious.

Chapter 4

Nonlinearity management for quadratic, cubic, and Bragg-grating solitons

4.1 The tandem model and quasi-phase-matching

The models considered in the previous chapters involved only $\chi^{(3)}$ (cubic, alias Kerr) nonlinearity. Quadratic ($\chi^{(2)}$) nonlinearities and, in particular, SHG (second-harmonic-generating) ones also play an important role in optics. The quadratic nonlinearity may be another natural ingredient of periodic heterogeneous nonlinear systems. In particular, a *tandem system*, which is a periodic concatenation of $\chi^{(2)}$-nonlinear and linear elements, was proposed by Torner [162] as a medium facilitating the creation of $\chi^{(2)}$ solitons in the temporal domain. The respective model is based on the SHG equations

$$iu_z + ic(z)u_\tau - \frac{1}{2}\beta(z)u_{\tau\tau} + \gamma(z)u^*v = 0,$$

$$2iv_z - ic(z)u_\tau + \frac{1}{2}\beta(z)v_{\tau\tau} + \frac{1}{2}\gamma(z)u^2 + q(z)v = 0 \qquad (4.1)$$

(cf. Eqs. (1.36) for the spatial-domain SHG model), which take into regard the walk-off c (alias group-velocity mismatch, GVM) between the FF and SH waves. The coefficients c, β and γ in Eqs. (4.1) take two different sets of values in periodically alternating intervals of z, so that one of them has $\gamma = 0$ (no nonlinearity) and much higher values of the GVD and GVM coefficients than the nonlinear segment. Generally, the strongly dispersive linear segments are (effectively) much shorter than the low-dispersive nonlinear ones.

Systematic numerical simulations performed in work [162] had demonstrated that robust oscillating solitons can readily self-trap in this model. In fact, the alternation of the parameters in the linear and nonlinear segments can facilitate the making of temporal $\chi^{(2)}$ solitons in comparison with the uniform waveguide (in particular, the tandem scheme may help to resolve the most difficult problem of compensating the usually large GVM between the FF and SH waves).

A well-known application of periodic heterogeneous structures in SHG media is the quasi-phase-matching (QPM) technique. This technique is implemented as periodic reversal of the orientation of electric-polarization domains in ferroelectric crystals, which are used as $\chi^{(2)}$ materials, or the periodic reversal of the orientation of poling, which provides for the $\chi^{(2)}$ nonlinearity in other settings. The reversal is periodic along the propagation direction, with a period $2\pi/Q$. This implies that the coefficient in front of the SHG terms in Eqs. (1.36) or 4.1) is a periodic function of z that can be expanded in a Fourier series, starting with the spatial harmonic $\exp(iQz)$. The extra wave vector Q, introduced this way, may be used to compensate the phase-velocity mismatch, if the latter is two large, which is often the case (for instance, the QPM period may be ~ 10 μm for the wavelength of light $\sim 1\ \mu$m, which means that Q may compensate the wave vector mismatch in the size of up to $\sim 10\%$ of the carrier wave vector itself).

Soliton propagation in the QPM medium was theoretically investigated [43], with a conclusion that the soliton may feel the action of an effective nonlinearity which includes not only $\chi^{(2)}$ terms, but also $\chi^{(3)}$ ones, induced through the cascading mechanism. In fact, the QPM technique can be used to *engineer* desirable effective nonlinearity, that may include rather exotic terms (for instance, with opposite signs of the effective SPM and XPM coefficients). A known example [27] of such a theoretically predicted engineered system is

$$iu_z + \frac{1}{2}u_{xx} + \eta_1 u^* v + \left(\gamma_2|u|^2 - \gamma_1|v|^2\right)u = 0,$$

$$iv_z + \frac{1}{4}v_{xx} + \eta_2 u^2 - 2\gamma_2|u|^2 v + qv = 0, \qquad (4.2)$$

where γ_1, γ_1 and η_2 are real coefficients selected by the the the engineering method. These equations were derived by averaging over the small-scale spatial modulation that represents the QPM.

Still more sophisticated systems can be created using a new theoretically and experimentally elaborated technique of the *quasi-periodic* (rather than periodic) QPM, see paper [68] and references therein. The latter technique makes it possible to provide for simultaneous matching of many sets of quadratically interacting waves, rather than of the single FF-SH set. However, the QPM models and their generalizations do not actually belong to the class of the periodic heterogeneous nonlinear systems, because averaged equations which describe the light propagation in $\chi^{(2)}$ media altered by means of the QPM technique, such as Eqs. (4.2), have constant coefficients.

4.2 Nonlinearity management: Integration of cubic and quadratic nonlinearities with dispersion management

A practically interesting application of the $\chi^{(2)}$ nonlinearity which leads to another example of the nonlinear periodic heterogeneous systems is the use of SHG modules for generating an effectively cubic (cascaded) nonlinearity with a negative Kerr coefficient. If periodically inserted in a long fiber-optic link, properly tuned SHG elements may provide for compensation of the nonlinear phase shift accumulated by RZ pulses passing long fiber spans, similar to how the DM provides for periodic compensation of the accumulated dispersion. This technique is known as *nonlinearity management* (NLM). Besides the application to the long-haul fiber-optic telecommunications, the utility of the technique was demonstrated in soliton-generating fiber-ring lasers [102], and for optical signal processing [33].

In terms of the optical telecommunications, NLM was proposed (in an abstract form, without specifying that the compensating elements would use cascaded $\chi^{(2)}$ nonlinearity) in work [139]. A full model, that includes the Kerr nonlinearity and DM in fiber spans, and SHG equations in the compensating modules, was developed in paper [54], which showed that the SHG modules provide not only for the nonlinearity compensation, but, what is quite important too, periodic *reshaping* of the pulses, and simultaneously help to suppress the earlier mentioned detrimental effect of the ISI (inter-symbol interference) between co-propagating pulses. Basic results demonstrating robust transmission of RZ pulses in this system are presented below, chiefly following work [54].

4.2.1 The model

The system is arranged in such a way that the carrier frequency of the optical signal propagating through the fiber link is, simultaneously, the fundamental frequency (FF) of the energy-conversion cascade in the $\chi^{(2)}$ module. Parameters of the module are selected so that the peak power of a given input signal corresponds to the complete conversion cascade, FF $\rightarrow$ SH $\rightarrow$ FF (SH stands for the second harmonic), therefore the portion of the pulse around its center passes the module wasting negligible energy to the generation of a residual SH component, that cannot couple into the fiber span and is therefore lost. However, for parts of the pulse corresponding to smaller values of the power, the same propagation length in the SHG module is essentially different from that corresponding to the complete cascade, therefore the energy loss is conspicuous farther from the pulse's center. This mechanism *reshapes* the pulse, chopping its wings off. The extra energy loss incurred by the reshaping is compensated by increase in the gain of optical amplifiers, which must be included in the full system in any case. The optimum arrangement has the SHG module placed immediately after the amplifier, which maximizes the nonlinear $\chi^{(2)}$ effects.

To elaborate the approach outlined above, one should consider equations describing the evolution of the amplitudes u and v of the FF and SH fields in the SHG medium, which are just the general equations (1.36) without the diffraction terms (the latter ones are irrelevant in the present context). In a notation slightly different from that adopted in Eqs. (1.36), the SHG equations are

$$\frac{du}{d\zeta} = -\frac{1}{2}i\kappa u^* v, \tag{4.3}$$

$$\frac{dv}{d\zeta} = -\frac{1}{2}i\kappa u^2 - iqv, \tag{4.4}$$

where ζ is the distance passed by the beam in the medium, and the asterisk stands for the complex conjugation, κ and q being the $\chi^{(2)}$-interaction coefficient and phase mismatch, respectively. Dissipative attenuation of the signal inside the SHG crystal and the GVM between the FF and SH signals are not included, as both are negligible for relevant propagation lengths. Nevertheless, the model does imply that the GVM must be zero (or very small), which turns out to be necessary for a different reason: as demonstrated in work [78], a condition tantamount to the zero GVM provides for equalization of the phase-velocity mismatch across channels in the WDM system.

The propagation of the signal in the fiber spans (with altering sign of the GVD coefficients, to provide for the dispersion compensation) obeys the ordinary NLS equation (1.48) for the field $u(z, \tau)$ in the DM system. In the analysis, the latter equation includes the linear fiber loss (the same as in Eq. (3.2)), and linear amplifiers are included too, as per Eq. (3.5). A peculiarity of the model is that the gain G of different amplifiers is not exactly the same; instead, it is adjusted at each node so as to have a fixed peak power of the signal entering the $\chi^{(2)}$ module right after the amplifier, for which the signal is shaped by the module in an optimum way, as described above. Typically, the thus selected values of the gain for the model with realistic parameters are scattered between 10 and 13 dB.

The reshaping of the pulse by the $\chi^{(2)}$ module is described by a numerical solution of Eqs. (4.3) and (4.4) (the equations themselves are integrable, but solutions are not available in an explicit analytical form). Taking the input pulse as

$$u_{\text{in}}(\tau) = \sqrt{p_{\text{in}}(\tau)} \exp\left[i\phi_{\text{in}}(\tau)\right], v_{\text{in}}(\tau) = 0, \tag{4.5}$$

the pulse exiting the SHG module becomes

$$u_{\text{out}}(\tau) = \sqrt{p_{\text{out}}(p_{\text{in}}(\tau))} \exp\left[i\left\{\phi_{\text{in}}(\tau) + \Delta\phi(p_{\text{in}}(\tau))\right\}\right], \tag{4.6}$$

where the power-transform function $p_{\text{out}}(p_{\text{in}})$ and the $\chi^{(2)}$ phase shift $\Delta\phi(p_{\text{in}})$ can be found in a numerical form [54].

4.2.2   Results: transmission of a single pulse

The initial pulse was launched, at the point $z = 0$, in the form of a chirp-free Gaussian (cf. the same waveform (3.30) used in the combined SSM-DM model),

$$u_0(\tau) = \sqrt{p_0} \exp\left(-\frac{\tau^2}{\tau_0^2}\right). \tag{4.7}$$

The propagation of the pulse is simulated in the following way. The initial pulse (4.7) was propagated over the distance corresponding to one span of the link, which was followed by its linear amplification according to Eq. (3.5) and shaping as per Eqs. (4.5) and (4.6). Then, the pulse was fed into the next span, with the opposite sign of the GVD, and so on. As mentioned above, the gain of each amplifier was adjusted (within the interval 10 dB -13 dB) to provide for a constant value of the pulse's peak power entering the $\chi^{(2)}$ shaper.

Without the DM, the shape of the pulse cannot be maintained for the propagation length exceeding 10 spans, and in most cases irreversible distortion of the pulse starts after 8 spans. The introduction of the DM as explained above (opposite signs of the

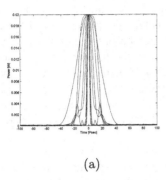
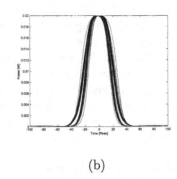

(a)   (b)

Figure 4.1: "Eye diagrams" illustrating the transmission of a single pulse in the system with the nonlinearity management provided for by the periodically installed $\chi^{(2)}$ modules. The diagrams are generated by juxtaposing the pulse profiles, $|u(\tau)|^2$, at the end of each span. (a) The system composed of 10 spans without the dispersion compensation; (b) the system composed of 50 spans with the zero average dispersion, i.e., with the full dispersion compensation (the DM strength corresponding to this example is $S = 0.9$).

GVD coefficients between adjacent spans) improves the situation drastically: at a fixed value of the GVD coefficient β_1 in the anomalous-dispersion span, variation of the GVD coefficient β_2 in the normal-dispersion one leads to a steep increase of the stable-transmission distance, almost by an instantaneous jump, from 8 spans to an indefinitely large number, when β_2 passes a relatively small critical value. Typical minimum values of the DM strength (defined as in Eq. (2.25)), which are necessary for the complete stabilization of the pulse, are quite small, $S_{\min} = 0.5 - 0.6$.

The drastic difference between the transmission of the pulses in the absence and presence of the DM is illustrated by "eye diagrams" for a pulse presented in Fig. 4.1. Diagrams of this type are frequently used in the analysis of pulse transmission in models of optical telecommunications. In the present case, they were generated by juxtaposing the pulse profiles, $|u(\tau)|^2$, at the end of each span passed. As is seen, in the NLM model without DM, the "eye" completely closes after the passage of 10 spans, which means that the pulse is not able to keep its shape, and is therefore unusable for the applications. On the contrary to that, in the system combining the DM and NLM, the eye remains completely open after 50 spans, i.e., the pulse is fully stable. The numerical analysis has demonstrated that, in the combined model, it is possible to secure the stable transmission through an indefinitely large distance, for the pulses with the width and amplitude taking values in very broad limits.

An essential aspect of the stability problem for the present system is robustness of the transmission against random variations of the initial peak power of the pulse. Indeed, the (effective) passage distance in the SHG modules, and the gain of each amplifier were selected so that the power at the center of pulses entering each module corresponded to the complete cascading cycle, FF $\rightarrow$ SH $\rightarrow$ FF. Noise-induced fluctuations in the initial peak power p_0 (see Eq. (4.7)) will violate this condition, and may therefore be potentially deleterious to the operation of the system. Investigation of the robustness of the system against disturbances of this type demonstrates that the operation regime is vulnerable against perturbations which make the initial peak power p_0 *smaller* (subtracting 1% from p_0 can essentially destabilize the pulse transmission). However, the transmission regime is fairly robust against perturbations that increase the peak power. For instance, in the case that was shown in Fig. 4.1, the pulse is destabilized only if its initial peak power is increased by more than 8% (if the initial perturbation exceeds this critical level, the pulse will get split into two after having passed ~ 10 spans).

It is relevant to note that the stability of the pulse against the increase of the initial power, and lack of stability against the decrease of the power, is an acceptable situation, as, in the case of random-noise perturbations, the powers of the unperturbed pulse and noise sum up, thus making the total power only larger than in the absence of perturbations.

4.2.3   Co-propagation of a pair of pulses

It was explained in the previous chapter that the ISI (inter-symbol interference) , i.e., gradual filling of the gap between adjacent pulses in a data-carrying stream, is an issue of great practical importance. In the present model, a generic result revealed by systematic simulations is that the stable co-propagation of pulses is possible through the distance in excess of 16 spans, although not much larger. The difference from the single-pulse case, where the stable transmission is possible for an indefinitely large number of spans, may be explained as follows: in the course of the propagation, each pulse emits small portions of radiation, which hit the other pulse. The accumulation of this perturbation eventually leads to strong distortion of the pulses. The situation could be improved if optical filters are added to the system (cf. the situation considered in the previous chapter for the soliton transmission in DM systems), as the filters absorb the radiation.

A necessary condition for the stability of the co-propagating pair of pulses is that the temporal separation between them must exceed a minimum value, $T_{\min}$. This characteristic is important as it sets a limit for the possible bit-rate capacity of the optical telecommunication link. Varying the DM strength within the (quite broad) interval of $0.5 < S < 5$, it was found that $T_{\min}$ attains a flat minimum in a subinterval $2.5 < S < 3.5$, as shown in Fig. 4.2. This result demonstrates that, as concerns the suppression of the interaction between pulses, the case of the *moderate DM* is an optimal one; recall that a similar conclusion has been made about the soliton interactions in the ordinary DM systems [171, 146].

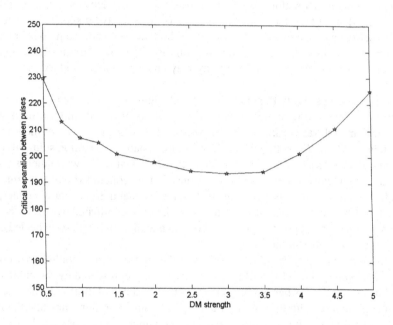

Figure 4.2: The minimum separation between the co-propagating pulses, necessary for the stable passage of (at least) 16 spans in the system combining the nonlinearity management and dispersion management, vs. the DM strength in the system with the zero average dispersion.

4.3 Nonlinearity management for Bragg-grating and nonlinear-Schrödinger solitons

4.3.1 Introduction to the problem

Gap solitons (GSs) in the model of the fiber Bragg grating (BG) with the cubic nonlinearity, based on Eqs. (1.27) and (1.28), and similar models constitute a separate class of solitons. A principal difference between them and the NLS solitons is that the latter ones exists as a result of the balance between the self-focusing SPM nonlinearity and anomalous temporal dispersion. If the nonlinearity is self-defocusing, while the dispersion remains anomalous, bright solitons do not exist. However, the sign of the nonlinearity does not matter in the BG model, as the effective dispersion (or diffraction, see below) induced by the grating includes both normal and anomalous branches, hence either of them will be able to support solitons. The latter circumstance suggests to consider a model where the nonlinearity may change its sign, and explore GSs in that case.

The simplest possibility to realize the sign-changing nonlinearity is to take it as a combination of cubic and quintic terms with opposite signs. This modification of the standard BG model was considered in work [20]. As well as in the case of the standard BG system (1.27), (1.28) with the cubic nonlinearity, stationary soliton solutions of its cubic-quintic counterpart were found in an exact analytical form, while their stability was studied by means of numerical simulations. It was concluded that the family of GSs in the modified system is drastically different from that in its standard counterpart: the family splits into two disjoint subfamilies, each being dominated by one of the two nonlinear terms of the opposite signs (in accordance with what might be expected), and a part of each subfamily is stable.

A different possibility to study the effect of the sign-changing nonlinearity on the GSs is to introduce a model with the nonlinearity being represented by the cubic term only, whose sign changes periodically as a function of the evolution variable, i.e., the NLM in combination with the BG. In the temporal domain, this implies that the nonlinearity must periodically change its sign in time, which is not a physically realistic assumption. However, the necessary arrangement can be implemented in the spatial domain, i.e., for stationary light beams propagating across a layered structure in a planar nonlinear waveguide.

4.3.2 Formulation of the model

According to what is said above, the model to be considered has the form

$$iu_z + iu_x + \gamma(z)\left(\frac{1}{2}|u|^2 + |v|^2\right)u + v = 0, \tag{4.8}$$

$$iv_z - iv_x + \gamma(z)\left(\frac{1}{2}|v|^2 + |u|^2\right)v + u = 0, \tag{4.9}$$

where z is the propagation distance, which plays the role of the evolution variable instead of time in Eqs. (1.27) and (1.28), and x is the transverse coordinate in the planar layered waveguide. The form of Eqs. (4.8) and (4.9) implies that the carrier wave vectors of the two waves, which are resonantly reflected into each other by BG, form equal angles with the z axis. The reflecting scores (or ribs) which form the BG with spacing h on the planar waveguide are oriented normally to the z axis, the Bragg-resonance condition taking the form of Eq. (1.29). The usual diffraction in the waveguide is neglected, as it is assumed that BG gives rise to a much stronger artificial diffraction. Although Eqs. (4.8) and (4.9) are z-dependent, they conserve the net power,

$$P \equiv \int_{-\infty}^{+\infty} \left(|u(x)|^2 dx + |v(x)|^2 \right) dx. \tag{4.10}$$

The layered structure of the waveguide assumes that the Kerr coefficient $\gamma(z)$ takes positive and negative values γ_+ and γ_- in alternating layers, cf. Eq. (3.29):

$$\gamma(z) = \begin{cases} \gamma_+, & \text{if } 0 < z < L_+ \\ \gamma_-, & \text{if } L_+ < z < L_+ + L_-, \end{cases} \tag{4.11}$$

which is repeated periodically with the period $L \equiv L_+ + L_-$. Using the scaling invariance of Eqs. (4.8) and (4.9), one may always impose the following normalization conditions:

$$L_+ + L_- \equiv 1, \; L_+\gamma_+ + L_-|\gamma_-| \equiv 1. \tag{4.12}$$

Thus, the model contains two irreducible control parameters, which may be selected as, e.g., L_+ and γ_+, while the other parameters can be found from Eqs. (4.12),

$$L_- = 1 - L_+, \; \gamma_- = -(1 - L_+\gamma_+)/(1 - L_+). \tag{4.13}$$

Note that corresponding average value of the Kerr coefficient is

$$\bar{\gamma} \equiv \frac{L_+\gamma_+ + L_-\gamma_-}{L_+ + L_-} = 2L_+\gamma_+ - 1. \tag{4.14}$$

Being interested in the sign-changing model, with $\gamma_- < 0$, the case of $L_+\gamma_+ \leq 1$ will be considered here, as it is equivalent to $\gamma_- \leq 0$ according to Eqs. (4.13).

Because the usual broad small-amplitude GS solitons (1.33) with $\theta \ll 1$ are asymptotically equivalent to broad NLS solitons, it is natural to consider, parallel to the model based on Eqs. (4.8), (4.9), also the spatial-domain NLS equation with the nonlinearity coefficient subjected to the same periodic modulation as in Eq. (4.11),

$$iu_z + \frac{1}{2}u_{xx} + \gamma(z)|u|^2 u = 0. \tag{4.15}$$

Comparing the results for the gap and NLS solitons in the two models will be quite helpful in realizing the generality of the conclusions presented below; besides that, the model (4.15) is of interest in its own right.

4.3.3   Stability diagram for Bragg-grating solitons

Unlike the NLS solitons, the variational approximation for the gap solitons is very complex even without NLM [113]. Therefore, one should rely on direct numerical simulations of Eqs. (4.8), (4.9), with $\gamma(z)$ defined as per Eqs. (4.11) and (4.13). The simulations used the initial configuration (at $z = 0$) in the form of the exact GS solution (1.33) for the uniform medium, parameterized by θ and taken at $t = 0$. The simulations were run for a fixed value of θ, while the model's control parameters γ_+ and L_+ were gradually varied, subject to the above-mentioned constraint $L_+\gamma_+ \leq 1$. Then, the same was done for other values of θ.

Numerical results identify a stability region for the solitons in the parameter plane (L_+, γ_+) which is shown in Fig. 4.3. The upper boundary of the stability region is $L_+\gamma_+ = 1$, which, as said above, limits the case considered here, as the local Kerr coefficient ceases to be sign-changing above this boundary. The upper boundary itself corresponds, as is seen from Eqs. (4.13), to a system in which nonlinear layers of the width L_+ alternate with linear ones (having $\gamma_- = 0$) of the width L_-. The simulations demonstrate that everywhere on this boundary, the solitons are stable, and they remain stable above the boundary, so by itself it is not a stability border. The left vertical boundary of the stability region in Fig. 4.3 at $L_+ = 0.2$ is not a real stability border – it only bounds a range for which the results are displayed (the range is $0.2 \leq L_+ < 1.0$).

The lower boundary of the stability region in Fig. 4.3 is rather close to a hyperbola, with the product $L_+\gamma_+$ along this boundary taking values between 0.65 and 0.70. For comparison, the dashed curve in Fig. 1 shows the hyperbola $L_+\gamma_+ = 1/2$, along which the average Kerr coefficient (4.14) exactly vanishes. The finite separation between the lower stability boundary and the dashed curve can be measured by the average value $\overline{\gamma}$ of the Kerr coefficient (4.14), the smallest value for $\overline{\gamma}$ found on the lower boundary being ≈ 0.3. Thus, stable solitons *are not possible* in a system where the average value of the Kerr coefficient is zero. This is, incidentally, a noteworthy difference from the DM system, where stable solitons are found in the case when the path-average dispersion exactly vanishes (see the previous chapter). This result is similar to one obtained in work [165], and presented in Chapter 7 below, for the (2+1)D (cylindrical) solitons in a bulk (3D) layered medium without BG: there too, a finite positive average value of the Kerr coefficient is necessary for the existence of any soliton, stable or unstable (see Fig. 7.1 and related text). It will be shown here that the same result is also true for the layered NLS model (4.15). On the other hand, a difference of the BG model from the NLS one is that the sign of the nonlinearity is not crucial, as explained above. Therefore, there exists another stability area in the region where the average Kerr coefficient is negative, i.e., $L_+\gamma_+ < 1/2$ (not shown in Fig. 4.3).

The formation of stable solitons in this system is accompanied by emission of radiation, and, sometimes, by generation of a small additional pulse, as shown in Fig. 4.4. The emission of radiation is conspicuous in the case when the stable soliton is close to the lower stability border (in terms of Fig. 4.3). On the other hand, in the unstable case the initial pulse completely decays into radiation.

The stability diagram displayed in Fig. 4.3 was obtained from simulations of Eqs. (4.8) and (4.9), starting from the initial configuration (1.33) with $\theta = 0.484 \cdot \pi$, which

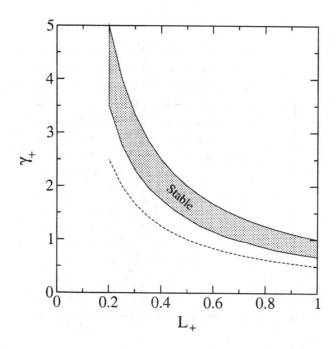

Figure 4.3: The stability diagram for the Bragg-grating solitons in the model with the nonlinearity management, based on Eqs. (4.8), (4.9) and (4.11), (4.13). The stability region is bounded by the lower solid curve, while the upper curve, the hyperbola $L_+\gamma_+ = 1$, is a border of the parametric area where the local Kerr coefficient periodically changes its sign. The dashed curve is the hyperbola $L_+\gamma_+ = 1/2$ along which the average value of nonlinearity is zero.

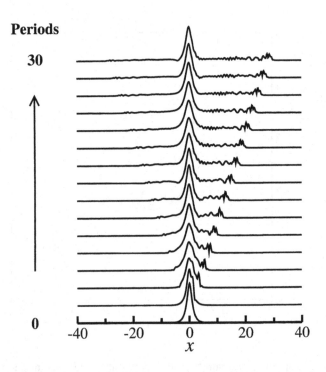

Figure 4.4: An example of the formation of a stable soliton in the Bragg-grating model with the sign-changing nonlinearity, for $L_+ = 0.5$ and $\gamma_+ = 1.7$. Only the u component is shown. An explanation to broken symmetry of the observed configuration (the small pulse has no mirror-image counterpart) is given in the text.

is close to the stability-limiting value $\theta_{cr} \approx 1.01 \cdot (\pi/2)$ of the standard model (1.27), (1.28) with constant coefficients. Additional simulations show that the stability region has virtually no sensitivity to the variation of θ: for example, decreasing θ from $0.484 \cdot \pi$ to $0.452 \cdot \pi$ (i.e., from $\arccos 0.05$ to $\arccos 0.15$; recall the relation $\omega = \cos \theta$ between the frequency of the usual BG soliton (1.33) and θ) produces absolutely no detectable change in the shape of the stability region.

4.3.4 Stability of solitons in the NLS equation with the periodic nonlinearity management

For the sake of comparison of the manifestations of the NLM in different models, it is relevant to generate a stability diagram for solitons in Eq. (4.15), with $\gamma(z)$ taken again as per Eqs. (4.11), (4.13). The simulations were run with the initial condition

$$u_0(x) = \eta \operatorname{sech}(\eta x), \qquad (4.16)$$

that would generate an exact soliton in the NLS equation with $\gamma \equiv 1$.

The result is that, for all the moderately narrow initial solitons (4.16), i.e., ones with η not too large, the respective stability diagram is almost indistinguishable from its counterpart in Fig. 4.3. A difference between the BG and NLS models becomes conspicuous if η in the initial pulse (4.16) is high. One may expect that, for very narrow solitons with large η, whose diffraction length $\sim 1/\eta^2$ is much smaller than the NLM period $L = 1$, the periodic change of the nonlinearity sign, as in Eq. (4.11), is a very strong perturbation that may destroy the soliton (cf. the situation for the SSM presented in the previous section, where solitons with the width ~ 1 cannot exist in the model with a very large modulation period L). Indeed, running the simulations of Eq. (4.15) with the initial condition (4.16), it was found that the stability region strongly shrinks, see an example for $\eta = 5$ in Fig. 4.5. Moreover, even if the evolution of the initial pulse (4.16) with large η results in the appearance of stable solitons, they were frequently produced by *splitting* of the initial pulse, as shown in Fig. 4.6. Recall that an initial chirpless pulse cannot split in the integrable NLS equation, but it does split in the SSM, provided that the energy of the pulse is large enough (Fig. 3.5).

4.3.5 Interactions between solitons and generation of moving solitons

Interactions between solitons in the nonlinearly-managed BG model based on Eqs. (4.8), (4.9), (4.11), and (4.13) were also studied in work [21]. In particular, two identical solitons with an initial phase difference $\Delta\phi$ attract each other if $\Delta\phi = 0$ (see Fig. 4.7(a)), and repel if $\Delta\phi = \pi$ (see Fig. 4.7(b)) or $\Delta\phi = \pi/2$ (not shown here). Figure 4.7 also demonstrates two other important features. First, it shows that stable "moving" solitons exists in the present model (in fact, in the spatial-domain model they are not moving, but are, actually, tilted spatial solitons in the (z, x) plane). Second, in the case when the two solitons initially attract each other, and hence temporarily merge into a "lump", as seen in 4.7(a), conspicuous *spontaneous symmetry breaking* is observed, and the outcome of the interaction is inelastic: an additional tilted pulse is generated,

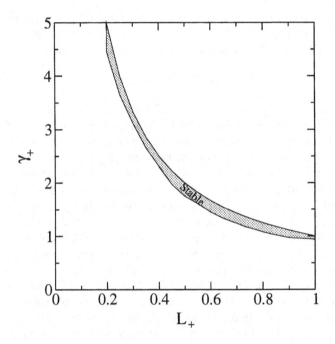

Figure 4.5: A narrow stability region in the nonlinear Schrödinger model (4.15) with the Kerr coefficient periodically changing its sign as per Eqs. (4.11) and (4.13), for the case when the initial pulse is given by Eq. (4.16) with $\eta = 5$.

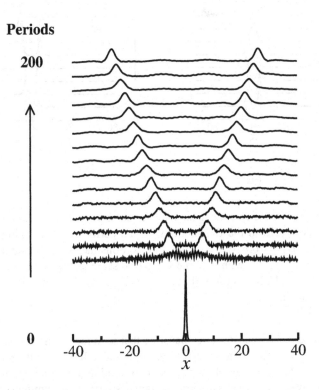

Figure 4.6: An example of splitting of the narrow pulse (4.16) with $\eta = 5$ in the model (4.15) into two secondary solitons with a smaller amplitude. The parameters of the model are $L_+ = 0.35$ and $\gamma_+ = 2.74$. The propagation distance shown in this figure is 200 modulation periods.

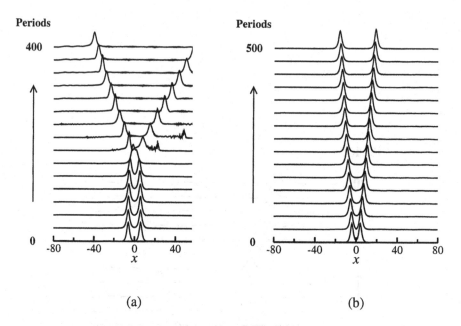

Figure 4.7: Interaction of two identical stable solitons with an initial phase difference $\Delta\phi$ and separation Δx in the BG model with the periodically changing sign of the nonlinearity, for $L_+ = 0.5$, $\gamma_+ = 1.94$. The solitons have been generated from the initial configuration (1.33) with $\theta \approx 0.484 \cdot \pi$ (the same as the one used in Fig. 4.4). (a) $\Delta\phi = 0$, $\Delta x = 12$ (attraction); (b) $\Delta\phi = \pi/2$, $\Delta x = 12$ (repulsion). Only the u component is shown.

along with some radiation. It seems plausible that a "lump", which was temporarily formed as a result of the attraction between the initial solitons, is subject to modulational instability, hence the amplification of small random numerical perturbations by the instability gives rise to the symmetry breaking. Note that the symmetry is preserved in the case of repulsion in Fig. 4.7(b), where no intermediate lump was formed.

Chapter 5

Resonant management of one-dimensional solitons in Bose-Einstein condensates

It was explained in the Introduction that the technique based on the Feshbach resonance (FR), which makes it possible to control the size and sign of the scattering length of atomic collisions in BEC, i.e., the coefficient in front of the cubic term in the corresponding GPE (Gross-Pitaevskii equation), has become a very important tool in the experiment [81, 148, 157], and also is a strong incentive for the development of theoretical analysis. Especially interesting is the possibility to control the nonlinearity coefficient by ac (time-periodic) magnetic field, which brings the nonlinearity management concept into the realm of BEC. In the 1D setting, this technique was developed under the name of the Feshbach-resonance management (FRM) in work [90]. The analysis was focused not on solitons proper, but rather on more general states whose localization was supported by the external parabolic trapping field (in the experiment, magnetic or optical trap is always necessary [141]). Additionally, the resonant action of the harmonic time modulation of the nonlinearity coefficient on fundamental and higher-order solitons in the NLS equation without external potential was recently studied in work [153]. Main results obtained for the FRM-driven localized nonsoliton and soliton states in the 1D setting are summarized in this chapter.

5.1 Periodic nonlinearity management in the one-dimensional Gross-Pitaevskii equation

It is known that the 3D Gross-Pitaevskii equation (GPE) (1.39) for BEC tightly confined in two transverse directions (x and y), and loosely confined by the parabolic potential $\Omega^2 x^2/2$ along the longitudinal axis x, can be reduced, by averaging in the transverse plane, to the 1D equation. In a normalized form, this effective GPE is

$$iu_t = -\frac{1}{2}u_{xx} + \frac{1}{2}\Omega^2 x^2 u + a(t)|u|^2 u, \qquad (5.1)$$

where $u(x,t)$ is the 1D wave function. Following work [90], the FRM-controled non-linearity coefficient, proportional to the scattering length, is modulated in time the same way as the GVD coefficient in the DM models is modulated as a function of the propagation distance z (see Eq. (1.49)),

$$a(t) = \begin{cases} a_1 > 0, & \text{if } 0 < t < T/2, \\ a_2 \lessgtr 0 & \text{if } T/2 < t < T, \end{cases} \qquad (5.2)$$

which is repeated with a period T. The value a_2 in Eq. (5.2) may be both positive and negative, but the most interesting case is one with $a_2 < 0$, when the nonlinearity coefficient periodically flips its sign. The modulation map (5.2) naturally defines the average ("dc") value of the nonlinear coefficient, and its "ac" part, as the amplitude of the periodic variation,

$$a_{\text{dc}} \equiv \frac{1}{2}(a_1 + a_2), \; a_{\text{ac}} \equiv \frac{1}{2}(a_1 - a_2). \qquad (5.3)$$

The main difference of this model from the NLM models considered in the previous chapter is an important role played by the parabolic trap (see below).

The most natural structure which may be expected to set in under the action of FRM, and in particular for small average values $\bar{a}$, is one oscillating between ground states that would exist at positive and negative constant values of a. In the former case ($a > 0$), this state is well approximated by the Thomas-Fermi (TF) wave function [141],

$$u_{\text{TF}} = \sqrt{\frac{2\mu - (\Omega x)^2}{2a}} e^{-i\mu t}, \qquad (5.4)$$

where μ is the chemical potential (determined by the number of atoms in the condensate). The TF approximation neglects the kinetic energy of the 1D motion of atoms in the condensate, i.e., the term u_{xx} in Eq. (5.1).

In the latter case, $a < 0$, a Gaussian wave function, i.e., the ground state of the quantum harmonic oscillator, is a natural approximation, unless a is too large (very strong nonlinearity). Numerical simulations corroborate this assumption: starting with the initial TF state, prepared for $a(0) = 1$ as per Eq. (5.4), simulations of Eq. (5.1) reveal persistent oscillations between the TF and Gaussian configurations, as shown in Fig. 5.1. For small values of a_1 and $|a_2|$, the oscillations are always regular (periodic). As a_1 and $-a_2$ increase to values ~ 1, more frequencies come into play, the oscillations become chaotic, which is accompanied by fragmentation of the wave function in space.

Overall description of dynamics of the FRM-driven 1D condensate is provided by a phase diagram which is displayed in the $(a_{\text{dc}}, a_{\text{ac}})$ plane in Fig. 5.2(a) (it may be relevant to compare this diagram with the diagram of dynamical states in the SSM model displayed above in Fig. 3.4). The "breather" (oscillating state) is stable beneath

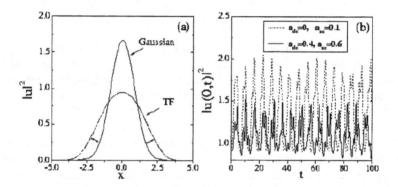

Figure 5.1: (a) Stable oscillations in the FRM-driven condensate in the weak parabolic trap with $\Omega = 0.002$, between the Thomas-Fermi and Gaussian configurations, in the case of $T = 2$, $a_{dc} = 0$, $a_{ac} = 0.1$, see definitions of the parameters in Eqs. (5.2) and (5.3). (b) Time evolution of the field's amplitude for $a_{dc} = 0$, $a_{ac} = 0.1$ (dotted line) and $a_{dc} = 0.4$, $a_{ac} = 0.6$ (solid line). In the former case, the oscillations are quasi-periodic, while in the latter case they are chaotic.

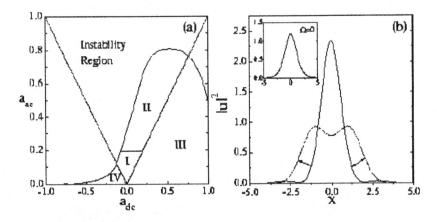

Figure 5.2: (a) The phase diagram for dynamical states of the FRM-driven condensate for $T = 2$. (b) Shape oscillations of the stable state close to the 2-soliton of the NLS type, for $a_{dc} = -0.4$ and $a_{dc} = 0.015$. The inset shows the fundamental-soliton stable state obtained instead of the 2-soliton if the trap is switched off, $\Omega = 0$.

the solid curve, featuring periodic and chaotic oscillations in areas I and II, respectively. In area III, the breather is closer to the TF state, which is quite natural, as the interaction is always repulsive in this case ($a_{dc} > a_{ac}$). Finally, in area IV, the breather strongly resembles the 2-soliton solution (1.15) of the self-focusing NLS equation, an example of which is shown in Fig. 5.2(b). It is relevant to mentioned that, in accordance with the fact that the 2-soliton is (weakly) unstable in free space, this breather demonstrates a completely different behavior in the absence of the trapping potential ($\Omega = 0$): it then sheds off $\approx 2\%$ of its norm with radiation, and reorganizes itself into an ordinary fundamental NLS soliton shown in the inset to Fig. 3.4(b).

Another type of the FRM-controlled dynamical state can be obtained by embedding a narrow *dark soliton* (DS) into the trapped condensate, thus creating a dip in the center of the breather. This can be done with the initial condition of the form $u = u_{TF}(x) \tanh x$ (recall the TF wave function is given by Eq. (5.4)). Then, a new stable state emerges, featuring drastically different dynamics: the central part of the condensate including the DS remains virtually static even as $a(t)$ takes negative values, as can be seen in Fig. 5.3 (cf. vibrations of the breather in Fig. 5.1), and only "wings" oscillate quasi-periodically between states with smaller and larger curvature (which may be regarded, respectively, as remnants of the former TF and Gaussian states). This soliton-like state may be viewed as a nonlinear counterpart of the first excited state of the quantum harmonic oscillator.

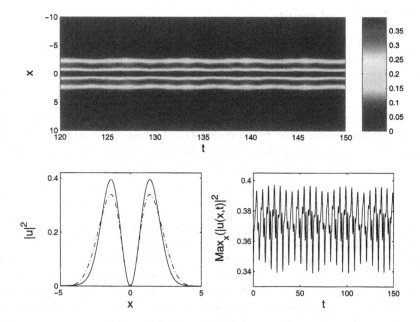

Figure 5.3: A dark soliton nested in the FRM-driven one-dimensional condensate with $T = 2$, $a_{\text{dc}} = 0$, and $a_{\text{ac}} = 0.1$. The top panel shows the evolution of the density. The left and right bottom panels display, respectively, the density profiles at $t = 139.6$ and $t = 142.8$ (by dashed-dotted and solid lines), and maximum density as a function of time.

5.2 Resonant splitting of higher-order solitons under the Feshbach-resonance management

5.2.1 The model

A very long effectively 1D condensate is described by the GPE without the trapping potential. The simplest possibility to introduce the FRM in this case is to add a small time-dependent harmonic ("ac", in terms of the previous section) term to a large constant ("dc") part of the coefficient in front of the cubic term, which corresponds to the attractive collisions between atoms in the condensate (negative scattering length). The corresponding version of GPE (5.1) is (in this section, the notation ϕ is adopted for the normalized 1D wave function)

$$i\phi_t + \frac{1}{2}\phi_{xx} + [1 + b\sin(\omega t)]|\phi|^2\phi = 0, \qquad (5.5)$$

where the amplitude b of the ac drive is small. Resonant splitting of higher-order solitons in this model was reported in work [153]. Note that Eq. (5.5) is similar to Eq. (4.15) dealt with in the previous chapter. However, the form of the periodic modulation of the nonlinearity coefficient was completely different there, and resonant effects were not considered in Eq. (4.15).

5.2.2 Numerical results

The n-soliton states in Eq. (5.5) with $b = 0$ are generated by the initial conditions
(1.14) with $\gamma = |\beta| = 1$, i.e.,

$$\phi_0(x) = N\eta \, \text{sech} \, (\eta(x - x_0)) , \tag{5.6}$$

where x_0 is the coordinate of the soliton's center. As the shape of the resulting state
oscillates with the period given by expression (1.16) (irrespective of the integer value
of n, for $N \geq 2$), a resonance may be expected if the driving frequency in Eq. (5.5) is
close to the frequency corresponding to the soliton period (1.16),

$$\omega_0 = 4\eta^2. \tag{5.7}$$

Figure 5.4 displays the evolution of the wave function generated by the initial condition (5.6) with $N = 2$ and $\eta = 1$, in the anticipated resonant case, with $\omega = \omega_0 = 4$
(pursuant to Eq. (5.7)) and the FRM-driving amplitude $b = 0.0005$. This very weak
resonant drive is sufficient to split the 2-soliton into two fundamental solitons. The
amplitudes of the splinters are very close to $\eta_1 = 3$ and $\eta_2 = 1$, exactly corresponding
to parameters of the fundamental-soliton constituents of the original 2-soliton with, as
per Eq. (1.17). Velocities of the splinters were measured to be

$$v_1 = 0.00197, \; v_2 = 0.0066, \tag{5.8}$$

respectively (with the ratio $v_1 : v_3 \approx 1 : 3$).

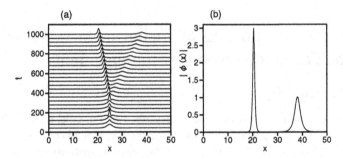

Figure 5.4: A typical example of splitting of a 2-soliton, generated by the initial condition (5.6) with $N = 2$ and $\eta = 1$, into an asymmetric pair of moving fundamental
solitons, under the action of the weak resonant FRM drive, with $\omega = 4$ and $b = 0.0005$.
(a) The evolution of $|\phi(x, t)|$. (b) The wave-function configuration at $t = 1000$.

Similar resonant splittings were observed for n-solitons with $n > 2$. In particular,
Fig. 5.5 shows this outcome for the initial configuration (5.6) with $N = 3$, $\eta = 0.5$,
$\omega = 1$ (which should be the resonant frequency, as per Eq. (5.7)), and $b = 0.0005$. This
time, the splitting gives rise to three moving fundamental solitons, whose amplitudes
are close to $A_1 = 2.5$, $A_2 = 1.5$, and $A_3 = 0.5$. These values precisely correspond to

the constituents of the original 3-soliton (with $A = 0.5$), as given by Eq. (1.18). The velocities of the three splinters are

$$v_1 = -0.00146, \ v_2 = 0.0732, \ v_3 = -0.0148, \qquad (5.9)$$

with ratios between them $v_1 : v_2 : v_3 \approx (-1) : 5 : (-10)$.

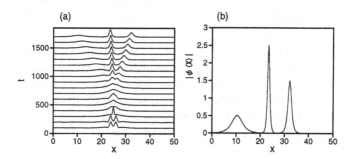

Figure 5.5: The same as in Fig. 5.4 for a 3-soliton, generated by the initial configuration (5.6) with $N = 3$ and $\eta = 0.5$. In this case, the forcing frequency and amplitude are $\omega = 1$ and $b = 0.0005$.

These results are summarized in Fig. 5.6 in the form of plots which show the minimum (*threshold*) value of the forcing amplitude b, necessary for the splitting of the 2- and 3-solitons as functions of the driving frequency ω. As seen from the figure, these dependences clearly have a resonant shape, with sharp minima at the frequency predicted by Eq. (5.7). Similar results were also obtained for the n-solitons with $n = 4$ and 5.

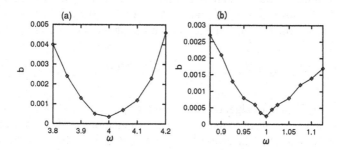

Figure 5.6: The minimum values of the amplitude b of the FRM driving term, necessary for the splitting of the 2-soliton (a) and 3-soliton (b),versus the driving frequency ω. The initial condition is taken in the form of Eq. (5.6) with, respectively, $N = 2$ and $\eta = 1$, or $N = 3$ and $\eta = 0.5$. In both cases, the sharp minimum exactly corresponds to the resonant frequency, as predicted by Eq. (5.7).

5.2.3   Analytical results

The amplitudes and velocities of the fundamental solitons, into which the higher-order ones split, can be predicted in an analytical from. It was already mentioned above that the amplitudes of the secondary solitons coincide with those which correspond to the constituents of the original n-soliton, as given by Eqs. (1.17) and (1.18). However, the velocities of the emerging fundamental solitons cannot be forecast this way, as, in terms of the IST technique applied to Eq. (5.5) with $b = 0$, which correctly predicts the amplitudes, the velocities must be zero.

Nevertheless, both the amplitudes and velocities of the final set of the solitons can be predicted in a different way, using the exact and nearly exact conservation laws of Eq. (5.5). These are two exact dynamical invariants, the norm (1.43) and momentum (1.10), and, in addition to them, three approximately conserved quantities are the Hamiltonian (1.8) and two higher-order expressions, (1.11) and (1.12) (they are conserved only approximately as the nonlinearity coefficient in Eq. (5.5) contains the small variable part).

In the case of the splitting of the 2-soliton (5.6) with the amplitude η, the exact conservation of the norm and approximate conservation of the Hamiltonian yield the following relations between η and the amplitudes $\eta_{1,2}$ of the emerging fundamental solitons (splinters): $4\eta = \eta_1 + \eta_2$, and $28\eta^3 \approx \eta_1^3 + \eta_2^3$ (the latter relation neglects small kinetic energy of the emerging solitons). These two relations immediately yield $\eta_1 = 3\eta$ and $\eta_2 = \eta$, which coincides with the the above-mentioned numerical results, as well as with the predictions based on the set of the 2-soliton's eigenvalues (1.17). Furthermore, the exact momentum conservation yields a relation involving the velocities $v_{1,2}$ of the secondary solitons, $\eta_1 v_1 + \eta_2 v_2 = 0$. With regard to the ratio $\eta_1/\eta_2 = 3$, this implies $v_1/v_2 = -1/3$, which is consistent with the numerical results (5.8), although the absolute values of the velocities cannot be predicted this way.

Similarly, in the case of the splitting of the 3-soliton, the exact conservation of N and approximate conservation of H and I_5 (see Eq. (1.12)) yield the relations (which again neglect small kinetic terms, in view of the smallness of the observed velocities)

$$9\eta = \eta_1 + \eta_2 + \eta_3, \, 153\eta^3 \approx \eta_1^3 + \eta_2^3 + \eta_3^3, \, 3369\eta^5 \approx \eta_1^5 + \eta_2^5 + \eta_3^5. \qquad (5.10)$$

A solution to this system of algebraic equations is $\eta_1 = 5\eta$, $\eta_2 = 3\eta$, $A_3 = \eta$, which are the same values that were found from direct simulations, and can be predicted as the IST eigenvalues (1.18). The conservation of P and I_4 gives rise to further relations,

$$\eta_1 v_1 + \eta_2 v_2 + \eta_3 v_3 = 0, \, (\eta_1 v_1^3 - \eta_1^3 v_1) + (\eta_2 v_2^3 - \eta_2^3 v_2) + (\eta_3 v_3^3 - \eta_3^3 v_3) = 0. \qquad (5.11)$$

If the velocities $v_{1,2}$ are small, it follows from here that $v_1/v_2 = -(\eta_2^3 - \eta_2\eta_3^2)/(\eta_1^3 - \eta_1\eta_3^2) = -1/5$, and $v_3/v_2 = -(\eta_3^3 - \eta_2\eta_1^2)/(\eta_3^3 - \eta_3\eta_1^2) = -2$. These ratios are consistent with the numerical results (5.9).

5.3 Resonant oscillations of a fundamental soliton in a periodically modulated trap

Besides the periodic modulation of the scattering length (that determines the nonlinearity coefficient in the GPE) by means of the FR, another experimentally feasible way to "ac-manage" dynamical states in the 1D condensate is to periodically modulate (in time) the strength of the parabolic trapping potential, as was proposed in papers [70, 71]. The corresponding normalized GPE in one dimension is

$$i\frac{\partial\psi}{\partial t} + \frac{1}{2}\frac{\partial^2\psi}{\partial x^2} + |\psi|^2\psi = [1 + \varepsilon\cos(\Omega t)]\, x^2\psi, \qquad (5.12)$$

where the interactions are assumed attractive, ε and ω being the amplitude and frequency of the modulation, cf. Eq. 5.5. In particular, the modulation of the trap may give rise to a resonance with harmonic oscillations that a soliton, as a quasi-particle, is expected to perform in the constant parabolic potential. In the latter connection, it is relevant to mention that *gap solitons*, supported by the additional potential in the form of an optical lattice in the BEC with repulsive interactions (see their description in the Introduction), have a *negative effective mass*, hence they may perform stable oscillations in an *anti-trapping* (inverted) parabolic potential, in the 1D [151] and 2D [152] settings alike.

5.3.1 Rapid periodic modulation

The 1D GPE with the trapping potential subjected to the time-periodic modulation was studied in detail in work [6] for the case of *rapid modulation* with a large frequency Ω. In that work, the trapping potential $U(x)$ could be more general than $U(x) = x^2$ in Eq. (5.12). In this case, a general solution may be looked for in the form of

$$\psi(x,t) = \Psi_{\text{slow}}(x,t) + \chi_{\text{rapid}}(x,t), \qquad (5.13)$$

where the parts $\chi_{\text{rapid}}(x,t)$ and $\Psi_{\text{slow}}(x,t)$ account for, respectively, rapid oscillations of the wave function due to the high-frequency modulation of the potential, and slow systematic evolution of the main part of the solution. The substitution of this expression in Eq. (5.12) and separating rapidly and slowly varying terms make it possible to find a small-amplitude (linearized) solution for $\chi_{\text{rapid}}(x,t)$, and then an effective equation for the slow evolution is derived by the method of averaging. As a result, with the use of an additional transformation of the slowly varying part of the solution (5.13),

$$\Phi^2(x,t) \equiv \left[1 + \frac{1}{2}\left(\frac{\varepsilon}{\Omega}\right)^2 U(x)\right]\Psi^2(x,t) \qquad (5.14)$$

(here, $U(x)$ is the constant part of the potential, e.g., $U(x) = x^2$, as said above), the eventual equation is cast in the form

$$i\frac{\partial\Phi}{\partial t} + \frac{1}{2}\frac{\partial^2\Phi}{\partial x^2} + |\Phi|^2\Phi = \left[U(x) + \left(\frac{\varepsilon}{2\Omega}\right)^2(U'(x))^2\right]\Phi. \qquad (5.15)$$

This equation contains a constant potential, which, however, is different from the original one, $U(x)$, as it includes an extra term $\sim (U'(x))^2$ (the prime, as usual, stands for the derivative) generated by the averaging procedure. Note that, if $U(x) = x^2$, the extra potential term is parabolic too. An example of the effective potential altered by the rapid modulation of the trap is shown in Fig. 5.7.

An interesting prediction of Eq. (5.15) is a possibility to stabilize the situation with the inverted parabolic potential, $U(x) = -Cx^2$, with a positive constant C. As it follows from Eq. (5.15), the full potential becomes normal (uninverted) under the condition $C (\varepsilon/\Omega)^2 > 1$.

5.3.2   Resonances in oscillations of a soliton in a periodically modulated trap

Analytical considerations

The high-frequency drive considered in the previous subsection cannot lead to resonances. On the other hand, periodic modulation applied at moderate frequencies, commensurate with eigenfrequencies of collective oscillations of the trapped condensate, may give rise to resonances. In particular, parametric resonances (PRs) induced by periodic modulation of the trap filled by the self-repulsive condensate in 1D, 2D, and 3D geometry, were studied in several works [71, 70, 7].

Another possibility is to investigate PRs in the motion of solitons in a 1D condensate with *attractive* nonlinearity, which was done in work [22]. Basic results are presented below, following that paper.

Assuming that the trap is effectively weak, the soliton may be approximated by the usual NLS ansatz (2.6), i.e., in the present notation,

$$\psi(x,t) = \eta \operatorname{sech}\left(\frac{x-\xi}{a}\right) \exp\left(i[\phi + w(x-\xi) + b(x-\xi)^2]\right), \qquad (5.16)$$

where η, a, ξ, ϕ, w, b are the real time-dependent amplitude, width, coordinate, wavenumber, and chirp of the soliton. The application of the standard VA technique leads to the following system of dynamical equations,

$$\ddot{a} = \frac{4}{\pi^2 a^3} - \frac{2N_s}{\pi^2 a^2} - 2\left[1 + \varepsilon \cos(\Omega t)\right] a, \qquad (5.17)$$

$$\ddot{\xi} = -2\left[1 + \varepsilon \cos(\Omega t)\right] \xi, \qquad (5.18)$$

where $N_s \equiv 2\eta^2 a$ is the conserved norm (proportional to the number of atoms in the condensate), and the overdot stands for d/dt. The other dynamical variables are given by relations $w = \dot{\xi}$ and $b = \dot{a}/(2a)$, cf. Eqs. (2.11)-(2.13).

Equation (5.17) is tantamount to one that was derived in the context of collective oscillations of 1D repulsive BEC held in the periodically modulated trapping potential [7].

Equations similar to (5.17) and (5.18) can also be obtained by means of the method of moments in a completely different problem, *viz.*, the evolution of optical beams in nonlinear graded-index fibers [144]. In that work, strong resonances in oscillations of

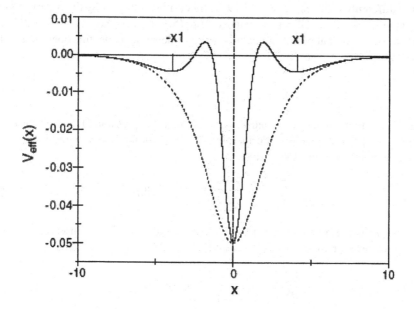

Figure 5.7: An example of the potential altered by the contribution from the rapid time modulation of the trapping potential, as per Eq. (5.15) (the example is taken from paper [6]). The original potential, shown by the dashed line, is $U(x) = -0.05\mathrm{sech}(0.6x)$, $\varepsilon = 180$, and $\Omega = 10$ (large ε helps to show the result in a clear form). The full (renormalized) potential, given by Eq. (5.15), is shown by the continuous line. Note that the renormalized potential acquires two local potential minima, at points $x = \pm x_1 \equiv 3.96$.

the beam's width were found in the case when the fiber's graded index is a piecewise-constant periodic function of the propagation coordinate, which is qualitatively similar to the periodic modulation of the trapping potential in the GPE.

The fact that equation (5.18) for the coordinate of the soliton's center is decoupled from the equation for the soliton's width is a general result, which is valid irrespective of the applicability of the VA. Indeed, precisely an equation in the form of (5.18) for the soliton's center-of-mass coordinate, which is defined as

$$\xi(t) \equiv \frac{1}{N} \int_{-\infty}^{+\infty} x|\psi(x,t)|^2 dx, \tag{5.19}$$

(recall N is the conserved soliton's norm), can be derived as an *exact* corollary of the GPE with the (time-dependent) parabolic potential. In fact, this is a manifestation of the *Ehrenfest theorem* in the present context (the validity of this theorem for the NLS with a parabolic potential was proved by Hasse [80]).

It is relevant to present here the derivation of Eq. (5.18) in the exact form. First, one should differentiate the expression (5.19) in time, substituting ψ_t by the full expression following from the 1D GPE (for instance, Eq. (5.12)). It is easy to see that, for the GPE with *any* external potential, including a time-dependent one, this operation yields an identity

$$\frac{d\xi}{dt} = \frac{P}{N}, \tag{5.20}$$

where P is the momentum defined as per the integral expression (1.10) (properly adjusted to the present notation). Further, the differentiation of the integral definition of P in time yields another exact result,

$$\frac{dP}{dt} = -\int_{-\infty}^{+\infty} U'(x)|\psi(x)|^2 dx, \tag{5.21}$$

where $U(x)$ is the potential in the GPE. Because the norm N is conserved independently, the insertion of $P = N d\xi/dt$ from Eq. (5.20) in Eq. (5.21) yields

$$\frac{d^2\xi}{dt^2} = -\frac{1}{N} \int_{-\infty}^{+\infty} U'(x)|\psi(x)|^2 dx. \tag{5.22}$$

Finally, substituting $U(x) = [1 + \varepsilon \cos(\Omega t)] x^2$ in Eq. (5.22) and once again taking into regard the definition (5.19), one arrives at Eq. (5.18).

Equation (5.18) is precisely the classical linear Mathieu equation (ME) [12]. It is commonly known that the ME gives rise to parametric resonances (PRs) when Ω is close to the values

$$\Omega_{PR}^{(n)} = 2\sqrt{2}/n, \tag{5.23}$$

$n = 1$ and $n > 1$ (n is integer) corresponding to the fundamental and higher-order resonances, respectively. In fact, Eq. (5.17) may be regarded as a nonlinear generalization of the ME, which also gives rise to PRs.

As concerns Eq. (5.17), which is not an exact one, but is only valid within the framework of the VA, it is relevant to mention that, in the low-density limit ($N_s \ll \pi^2 a^2/2$), the second term on the right-hand side of the equation may be dropped. The respective simplified equation is equivalent to an *exact* equation for the width, which was derived in paper [71] (without the use of the VA or other approximations) from the GPE in two dimensions (no such exact equation is available in 1D or 3D case) with the repulsive nonlinearity and parabolic trapping potential. It is known that solutions of the latter equation can be expressed, by means of an exact transformation, in terms of solutions of the linear ME. Therefore, in the limit when the underlying GPE goes over into the linear Schrödinger equation, which corresponds to $N_s \to 0$, the PRs in Eq. (5.17) are exactly the same as in Eq. (5.18). However, Eq. (5.17) cannot be reduced to the linear ME in the general case (for finite N_s).

Numerical results

The trivial solution of the Mathieu equation (5.18), $\xi \equiv 0$, loses its stability in certain zones in the parameter plane (Ω, ε), close to the PR points (5.23) [12]. In that case, the solution features oscillations with a permanently growing amplitude. On the other hand, solutions of Eq. (5.17), which is a nonlinear generalization of the ME, are always oscillatory ones (obviously, this equation has no trivial solution), and it may be expected that, also close to the points (5.23), a periodic solution to the latter equation will develop its own instability, that will manifest itself too in unlimited growth of the amplitude of the oscillations ("swinging"). However, a difference from the linear ME should be in the swinging period: the variable $a(t)$ in Eq. (5.17) cannot pass through zero (the width must always be positive), and in the case of large-amplitude oscillations of $a(t)$, it will suddenly bounce back from a vicinity of $a = 0$, instead of crossing into the unphysical region of $a < 0$. This implies that the swinging period in Eq. (5.17) must be half of that in the linear equation 5.18.

This expectation is corroborated by simulations of Eq. (5.17). Actually, the PR-induced instability is always identified in simulations as permanent growth of the amplitude of oscillations. This definition makes the onset of the instability in Eqs. (5.18) and (5.17) identical, as for large a (which corresponds to large amplitudes of the oscillations) the two equations are nearly identical, except for the above-mention peculiarity of Eq. (5.17), that $a(t)$ must bounce back from $a = 0$. Thus, the *double parametric resonance* is expected in the system.

An issue of obvious interest is to explore manifestations of the double PR, which was predicted within the framework of the approximation based on the ODE system (5.18) and (5.17) (recall only the former equation is an exact one), in PDE simulations of the full model (5.12). The double PR was indeed observed in direct simulations, in the form of the growth of the amplitude of the oscillatory motion of the soliton (*external instability*), concomitant to permanent increase of the amplitude of the soliton's intrinsic vibrations (*intrinsic instability*). To identify the latter effect, $\xi(t)$ was extracted from results of the PDE simulations according to Eq. (5.19), and $a(t)$ – as per a natural definition,

$$a^2(t) \equiv N^{-1} \int_{-\infty}^{+\infty} [x - \xi(t)]^2 |\psi(x,t)|^2 \, dx \tag{5.24}$$

(recall that N is the norm of the solution). An example of the dual instability, generated by the double PR, is displayed in Fig. 5.8. Note that, in accordance with what should be expected (as explained above), the swinging period of $\xi(t)$ is indeed seen to be twice that of $a(t)$.

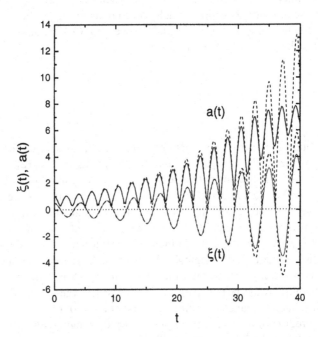

Figure 5.8: An example of the double parametric resonance in oscillations of a soliton in Eq. (5.12). The harmonic trap is periodically modulated at the fundamental-resonance frequency $\Omega = 2\sqrt{2}$ (see Eq. (5.23)), with the amplitude $\varepsilon = 0.2$. Simulations were performed with the initial condition $\psi_0(x) = \text{sech}(x-0.5)$ (the initial offset of the soliton from the trap's center, $x = 0$, leads to the oscillations). The corresponding results for $\xi(t)$ and $a(t)$, shown by continuous curves, were generated by means of Eqs. (5.19) and (5.24). They are compared with results of simulations of ODEs (5.18) and (5.17), which are shown by dashed curves.

In this figure, one can see some difference between the oscillation law for $\xi(t)$ as found from the direct simulations of Eq. (5.12), and from the numerical integration of ODE (5.18). An explanation to this is that the soliton under periodic perturbation emits linear waves which are eliminated by absorbers set at edges of the integration domain. As a result, the norm of the soliton slowly decreases, while the above derivation of Eq. (5.18) presumed a constant norm. The loss of the norm also explains strong deviation of the oscillations of $a(t)$ from the prediction of Eq. (5.17), observed in Fig. 5.8 at a late stage of the evolution. As concerns correspondence to the experiment, the absorbers emulate evaporation of atoms from a finite-size trap, which is a real physical effect.

Results of systematic direct simulations of Eq. (5.12) are summarized in the map of instability zones in the parametric plane (Ω, ε), which is displayed in Fig. 5.9. Zones shown in this figure reveal three separate PRs, *viz.*, the fundamental one at $\Omega = 2.82$, obviously corresponding to $n = 1$ in Eq. (5.23), and two higher-order PRs, at $\Omega = 1.41$ and $\Omega = 0.94$, which correspond to $n = 2$ and $n = 3$, respectively. The instability growth rate rapidly decreases for higher-order resonances, which explains why the PRs corresponding to $n > 3$ cannot be easily spotted in simulations running for a finite time. This also explains the fact that the instability "tongues" corresponding to the PRs with $n = 2$ and 3 do not extend to very small values of ε in Fig. 5.9.

Borders of the intrinsic-instability zones in Fig. 5.9, are, generally, close to the borders of the external instability (recall the latter are strictly tantamount to the instability borders in the parametric plane of the ordinary ME), except for a notable upward shift of all the intrinsic-instability zones, including the one corresponding to the fundamental PR at $\Omega = 2.82$. A reason for the shift is the above-mentioned radiation loss, which may be interpreted as effective dissipation. Accordingly, a more accurate approximation could be provided by a weakly damped nonlinear ME, instead of Eq. (5.17). It is known that weak friction indeed shifts the instability zones of the ME upward in ε, without affecting the resonant frequencies (in the first approximation) [147].

Finally, it is relevant to stress that, although Fig. 5.9 displays what was defined as instability zones, the soliton, even after the amplitude of its intrinsic vibrations starts to grow, does *not* feature self-destruction, remaining a coherent, although unsteady, object. Eventually, it gets destroyed, but only when it hits the absorbers at edges of the integration domain.

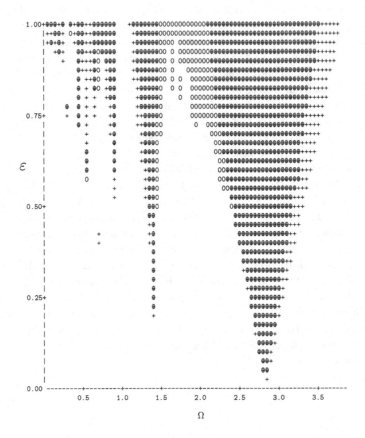

Figure 5.9: Instability zones, as found from direct simulations of the Gross-Pitaevskii equation (5.12) with the periodically modulated parabolic trap and self-attractive non-linearity. In the area covered by open circles, the oscillating soliton develops the intrinsic instability, in the form of a growing amplitude of the internal vibrations. Crosses cover regions where the soliton demonstrates the external instability (indefinite growth of the amplitude of oscillations of its center). The double parametric resonance occurs where both areas overlap.

Chapter 6

Management for channel solitons: a waveguiding-antiwaveguiding system

6.1 Introduction to the topic

The simplest way to stabilize and guide spatial optical solitons is to use *channels* for them, in the form of waveguides (WGs), i.e., stripes in nonlinear planar waveguides with a locally enhanced refractive index (RI). Multichannel systems are fabricated as sets of parallel waveguides. *Antiwaveguides* (AWGs) are structures with a reverse, relative to the ordinary WG, distribution of the linear RI between the core and cladding, see paper [74] and references therein. In the linear approximation, the light is ejected from the AWG's core into the cladding; however, a beam can be trapped inside AWG by the Kerr nonlinearity, provided that the beams's power exceeds a certain threshold value. An advantage of AWGs is that they may have very small cross sizes of both the core and trapped light beam, down to the order of the wavelength [74].

The antiwaveguided propagation is always unstable; however, the instability may be mild enough, being suitable for the design of all-optical multichannel switching schemes [74, 75]. In particular, an effective way to control the instability, initiating it at a point where the switching is required, is provided by the so-called "hot spot" (HS) [115], i.e., a spot attracting the propagating signal, which can be created, via the XPM, by a control laser beam shone perpendicular to the guiding structure and focused on the necessary spot off the AWG's axis (see Fig. 6.1 below).

In work [85], a new type of a nonlinear guiding structure was proposed, which, sharing with the usual AWGs their potential for switching applications, may be strongly stabilized, so that the length of stable propagation can be made, as a matter of fact, as long as required. The structure is an *alternate waveguide*, built as a periodic concate-

nation of AWG and WG sections. Clearly, it belongs to the class of the periodic heterogeneous nonlinear systems. In fact, the very concept of this class was for the first time
put forward in the same paper [85] where the alternate waveguides were proposed. A
unifying principle for the class, which was formulated in that paper, and is developed
in the present book, is strong stability of solitons in such systems, contrary to an *a
priori* expectation that coherent pulses would be quickly destroyed traveling through
strongly heterogeneous structures. This chapter presents main analytical and numerical results for *channeled* spatial solitons trapped in alternate waveguides, following the
paper [85].

6.2   The alternate waveguiding-antiwaveguiding struc-
ture

The alternate waveguide is a channel structure with equal Kerr coefficients in the core
and cladding, and a periodic RI modulation, $n \equiv n_0 + \delta n(z, x)$, along the propagation
axis (z) as schematically shown in Fig. 6.1:

$$\delta n(z) \equiv n(x = 0, z) - n(x = \infty, z) = \begin{cases} \delta n_+ > 0, & \text{in WG segments} \\ \delta n_- < 0, & \text{in AWG segments} \end{cases} \quad (6.1)$$

(the values of δn_+ and $|\delta n_-|$ are, in the general case, different). It is assumed that
the RI in the cladding is constant, $n(x = \infty, z) \equiv n_0$, so that the modulation of n is
limited to the core. A typical range of physically realistic values of the RI change in
waveguides is $|\delta n| \leq 0.01$.

In the usual paraxial approximation, the evolution of the local amplitude of the
electromagnetic wave, $\Psi(x, z)$, obeys the spatial NLS equation, whose normalized
form is (cf. Eq. (1.26))

$$i\frac{\partial \Psi}{\partial z} + \frac{\partial^2 \Psi}{\partial x^2} = [E + U(x, z)] \Psi - |\Psi|^2 \Psi, \quad (6.2)$$

where E is an effective propagation parameter (wavenumber), and $U(x, z) \sim n_0 \delta n(x, z)$
is an effective channel potential. Sections of the system with $U < 0$ (potential wells)
and with $U > 0$ (potential hills) correspond, respectively, to the WG and AWG segments. Detailed derivation of Eq. (6.2) from the full propagation equation can be found
in paper [85].

As is known, the RI profile produced by the diffusion technology which is used
for the fabrication of the WG/AWG core may be approximated by the function erf.
Therefore, the refractive index distribution in the AWG and WG parts of the alternate
waveguide may be assumed to be

$$n(x, z) = n_0 + \delta n(z) \cdot f(x), \ f(x) \equiv \frac{1}{2}\left[\text{erf}\left(\frac{x_0 + x}{D}\right) + \text{erf}\left(\frac{x_0 - x}{D}\right)\right], \quad (6.3)$$

where $\delta n(z)$ is defined in Eq. (6.1), x_0 is the effective half-width of the core, which
is 1 in the notation adopted above, and D is a fabrication (diffusion) parameter that

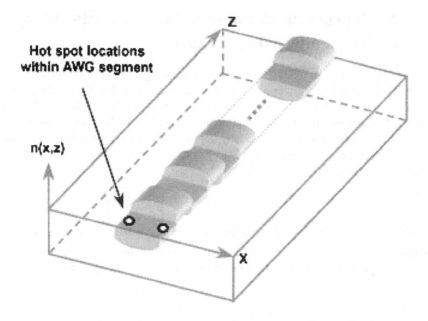

Figure 6.1: Schematic of the refractive index distribution in the waveguiding-antiwaveguiding alternate structure.

determines the eventual effective width of the guiding structure. Both WG and AWG segments of the alternate waveguide shown in Fig. 6.1 are assumed to have equal values of D. Finally, the effective potential corresponding to this channel structure, being proportional to $\delta n(x, z)$, is

$$U(x, z) = -A(z)f(x), \tag{6.4}$$

where the function $f(x)$ is the same as in Eq. (6.3), and the amplitude $A(z)$ periodically jumps between negative and positive values, as is typical to the periodic heterogeneous systems (cf. Eqs. (1.49), (3.29), and (4.11)),

$$A(z) \equiv \begin{cases} A_+ > 0, & \text{in WG segments} \\ A_- < 0, & \text{in AWG segments} \end{cases} . \tag{6.5}$$

6.3   Analytical consideration of a spatial soliton trapped in a weak alternate structure

Exact analytical solutions to Eq. (6.2) with the potential given by the expressions (6.5) and (6.3) are not available even for a stationary beam described by a real function $\Psi(x)$ in a uniform (WG or AWG) system. Nevertheless, stability of a spatial soliton (beam) propagating in the alternate structure with a *small* strength A can be investigated analytically. Indeed, in this case one may apply the perturbation theory which treats the solitary beam as a quasiparticle [104]. To this end, a beam solution to Eq. (6.2) with the potential (6.4), is sought for as

$$\Psi(x, z) = \exp\left[iqx + i\phi(z)\right] \Psi_0(x - \xi(z)), \tag{6.6}$$

where $\Psi_0(x) = \sqrt{2E}\mathrm{sech}\left(\sqrt{E}x\right)$ is the shape of the spatial soliton in the uniform medium, $\phi(z)$ is its phase, and $\xi(z)$ is a small off-center deflection of the soliton. The dynamical equation produced by the perturbation theory at the lowest order for ansatz (6.6) is:

$$\frac{d^2\xi}{dz^2} = -\frac{A(z)}{M}\frac{dW}{d\xi}, \tag{6.7}$$

where the effective mass and quasi-particle's potential are

$$M = \int_{-\infty}^{+\infty} \Psi_0^2(x)dx \equiv 4\sqrt{E}, \quad W(\xi) = \int_{-\infty}^{+\infty} \left[\Psi_0(x - \xi)\right]^2 f(x)dx \tag{6.8}$$

(a separate equation for the phase $\phi(z)$ is not displayed here, as it is not used in the analysis).

To study the stability of the beam guided in the channel, it is sufficient to linearize Eq. (6.7) in $\xi(z)$, which yields

$$\frac{d^2\xi}{dz^2} = -\frac{U_0'}{M}A(z)\xi \equiv -\omega^2(z)\,\xi. \tag{6.9}$$

Here, $U_0' \equiv (dU/d\xi)|_{\xi=0}$, and, with regard to Eq. (6.5),

$$\omega^2(z) = \begin{cases} (U_0'/M)\,A_+ \equiv \omega_+^2, & \text{in WG segments} \\ -(U_0'/M)\,|A_-| \equiv -\omega_-^2, & \text{in AWG segments} \end{cases} \tag{6.10}$$

Equation (6.9) describes a concatenation of stable oscillations with the frequency ω_+ in the WG segments, and unstable motion with the instability growth rates ω_- in the AWG ones. To predict the stability or instability of the channeled beam, it is necessary to find an explicit solution and conclude whether it is growing or remains confined, on the average, as the beam passes a large number of the WG-AWG cells. This implies solving Eq. (6.9) inside each interval where $\omega(z)$ is constant, and then matching the solutions, maintaining the continuity of $\xi(z)$ and $d\xi/dz$ across junctions between different segments.

The solutions inside the WG and AWG segments have, respectively, the form

$$\xi_{\text{WG}}(z) = a_+ \cos(\omega_+ z) + b_+ \sin(\omega_+ z), \tag{6.11}$$
$$\xi_{\text{AWG}}(z) = a_- \cosh(\omega_+ z) + b_- \sinh(\omega_+ z), \tag{6.12}$$

with arbitrary constants $a_\pm$ and $b_\pm$. It is a straightforward algebraic exercise to find relations between the two sets of the constants which follow from the conditions of the continuity of $\xi(z)$ and $d\xi/dz$. If a WG segment is followed by an AWG one, they are

$$a_- = a_+ \cos(\omega_+ L_+) + b_+ \sin(\omega_+ L_+),$$
$$b_- = (\omega_+/\omega_-)\,[-a_+ \sin(\omega_+ L_+) + b_+ \cos(\omega_+ L_+)], \tag{6.13}$$

and in the opposite case

$$a_+ = a_- \cosh(\omega_- L_-) + b_- \sinh(\omega_- L_-),$$
$$b_+ = (\omega_-/\omega_+)\,[a_- \sinh(\omega_- L_-) + b_- \cosh(\omega_- L_-)], \tag{6.14}$$

where L_+ and L_- are the lengths of the WG and AWG segments, respectively (their ratio is frequently called a *duty cycle*).

The product of the two linear transformations (6.13) and (6.14) yields a *map* which describes the transformation of the perturbation amplitudes $a_\pm$ and $b_\pm$ after the passage of a full cell of the alternate structure. It takes the form of a matrix,

$$\begin{pmatrix} \cos\chi_+ \cosh\chi_- - \frac{\omega_+}{\omega_-}\sin\chi_+\sinh\chi_- & \sin\chi_+\cosh\chi_- + \frac{\omega_+}{\omega_-}\cos\chi_+\sinh\chi_- \\ -\sin\chi_+\cosh\chi_- + \frac{\omega_-}{\omega_+}\cos\chi_+\sinh\chi_- & \cos\chi_+\cosh\chi_- + \frac{\omega_-}{\omega_+}\sin\chi_+\sinh\chi_- \end{pmatrix},$$

where $\chi_\pm \equiv \omega_\pm L_\pm$. The stability of the trapped beam is determined by the size of *multiplicators* of the map, i.e., eigenvalues $\mu_{1,2}$ of the matrix, the stability condition being $|\mu_{1,2}| \leq 1$, which must hold for both eigenvalues simultaneously. As the system under consideration is a conservative one, the only actual possibility for the stable propagation is that both $|\mu_{1,2}|$ are *exactly* equal to 1.

An elementary calculation of the eigenvalues yields

$$\mu_{1,2} = \tau/2 \pm \sqrt{\tau^2/4 - 1}, \tag{6.15}$$

where τ is the trace of the matrix,

$$\tau = 2\cos(\omega_+ L_+)\cosh(\omega_- L_-) + (\omega_-^2 - \omega_+^2)(\omega_+\omega_-)^{-1}\sin(\omega_+ L_+)\sinh(\omega_- L_-).$$

$$(6.16)$$

As it follows from Eqs. (6.15) and (6.16), the stability conditions $|\mu_{1,2}| = 1$ are met if $|\tau| \leq 2$, or, in an explicit form,

$$\left|\cos(\omega_+ L_+)\cosh(\omega_- L_-) - \frac{\omega_+^2 - \omega_-^2}{2\omega_+\omega_-}\sin(\omega_+ L_+)\sinh(\omega_- L_-)\right| \leq 1. \quad (6.17)$$

This inequality represents the full stability region in an explicit form, as predicted by the analytical approximation. Note that the condition (6.17) is trivially satisfied in the case of the uniform WG, $L_- = 0$, and it is definitely not satisfied in the opposite case of the uniform AWG, $L_+ = 0$.

It is easy to find a minimum value $(L_+)_{\min}$ of the WG segment which is necessary to achieve the stabilization at given values of the parameters L_- and $\omega_\pm$: as it follows from Eq. (6.17),

$$\omega_+(L_+)_{\min} = \arccos\left(\frac{1}{\sqrt{\cosh^2(\omega_- L_-) + \Omega^2\sinh^2(\omega_- L_-)}}\right)$$

$$- \arctan(\Omega\tanh(\omega_- L_-)), \quad (6.18)$$

where $\Omega \equiv (\omega_+^2 - \omega_-^2)/(2\omega_+\omega_-)$. It is easy to show that the expression (6.18) is always positive. Note that it remains finite in the limit of $L_- \to \infty$, which is explained by a possibility to find a special value of the parameter $\omega_+ L_+$ such that a WG segment inserted between two AWG segments mixes the eigenmodes so that the one corresponding to the unstable eigenvalue $+\omega_-$ in the AWG section preceding the WG one will go over into an eigenmode corresponding to the stable eigenvalue $-\omega_-$ in the next AWG section. In fact, the mixing of the stable and unstable AWG eigenmodes by the WG segments is a mechanism which makes the stable channeling of the beam in the alternate structure possible.

A noteworthy property of the stability condition (6.17) is that the stabilization of the trapped beam does *not* monotonically enhance with the increase of the duty-cycle ratio L_+/L_- of the WG and AWG segments. For instance, in the case of $\omega_+^2 = \omega_-^2$ the inequality (6.17) takes the form $|\cos(\omega_+ L_+)| \leq \mathrm{sech}(\omega_- L_-)$. If one increases L_+, keeping L_- fixed, the latter condition is met in intervals

$$\arccos[\mathrm{sech}(\omega_- L_-)] + 2\pi n \leq \omega_+ L_+ \leq -\arccos[\mathrm{sech}(\omega_- L_-)] + 2\pi(n+1),$$

$$(6.19)$$

$n = 0, 1, 2, ...$, and it is *not* met in gaps between the intervals. An implication of this is that, even if the length of the AWG segment is very small, there are some values of the ratio L_+/L_-, which may be arbitrarily large, around which the trapped beam is unstable. The non-monotonic dependence of the stability on the duty cycle is confirmed by numerical results displayed in the following section.

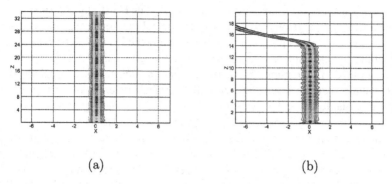

(a)   (b)

Figure 6.2: (a) Indefinitely long stable propagation of the spatial soliton in the waveguiding-antiwaveguiding system is shown by means of contour plots. The lengths of the respective segments are $L_+ = L_- = 2$, and the strengths are $A_+ = 5, A_- = -3$ (see Eq. (6.5)). For comparison, panel (b) shows one of the least unstable examples of the propagation of the beam in the uniform antiwaveguide, with $A = -5$, and parameters of the input pulse (6.20) $C^2 = 5, \sigma = 1$. In both case, the propagation parameter is $E = 2$.

6.4 Numerical results

6.4.1 Beam propagation in the alternate structure

The analytical results presented in the previous section strongly suggest that trapped beams may be stable in the alternate structure, but definite results concerning the stability can only be obtained from direct simulations. A realistic input profile is one with a Gaussian shape,

$$\Psi(z = 0) = C \exp[-(x/\sigma)^2]. \tag{6.20}$$

Simulations demonstrate that the strongest result, i.e., indefinitely long stable propagation with the smallest length-share of the inserted stabilized WG segments, could be achieved with equal lengths of the AWG and WG sections (50% duty cycle), both being on the order of the core width. An estimate shows that the corresponding lengths of the AWG and WG sections in physical units is $\sim 25\lambda$, where λ is the carrier wavelength [85]. A typical example of the thus stabilized propagation in the alternate waveguide is shown in Fig. 6.2; for comparison, unstable propagation in the uniform AWG is also shown (in fact, the later case is one of the least unstable ones possible in the uniform AWG).

6.4.2 Switching of beams by the hot spot

The above-mentioned "hot spot" (HS) can provide switching by pushing the beam from the core into the cladding at a necessary value of the propagation distance. In the simulations, the HS was approximated by a small increase of the refractive index, $(10^{-2} - 10^{-3})|\delta n|$, in a localized region of the AWG section, whose size was 1×1 in the normalized units, and δn is the same as in Eq. (6.3). The HS strength corresponding to these values of the parameters is sufficient to control the switching; in real-world units, this implies that the net power of the beam creating the HS should be between 1% and 0.1% of the signal-beam's power (this power, although small, is much higher than the fluctuation level, i.e., random fluctuations cannot switch the beam accidentally). Detailed simulations demonstrate that a virtually identical switching effect is produced by various particular forms of the distribution of the refractive-index perturbation inside the HS, provided that the net perturbation, integrated over the HS, is fixed.

The simulated switching, in the form of the controllable deflection of the beam from the guiding core into the cladding (where, in fact, it may be easily captured by another guiding channel, as it was described in detail in Ref. [75]) under the action of the HS is displayed in the Fig. 6.3. In this example, the lengths of the AWG and WG sections are 4 and 1, respectively. The corresponding duty cycle is taken small, i.e., not promising a very large length of the stable propagation, as very long propagation is not needed for the switching.

In Fig. 6.3, many particular realizations of the switching to the left and right at different values of z are juxtaposed. Each time, the switching position is selected by placing the HS at an appropriate place, as shown above in Fig. 6.1. Note that the HS in Fig. 6.3 is always set near the end of an AWG section, close to the beginning of the following WG one, as it was found that it acted most efficiently at this position, so that its strength might be minimized. It is easy to understand the highest sensitivity of the beam to the action of the HS at this place, as the potential instability accumulates to a maximum at the end of the AWG segment.

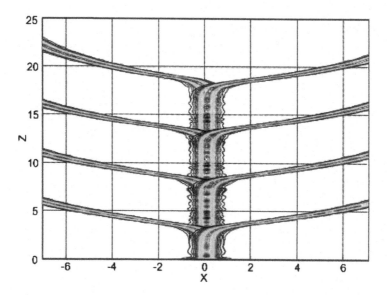

Figure 6.3: Juxtaposition of switchings induced by the "hot spot" applied on the left or right side off the axis to different anti-waveguiding sections in the alternate waveguide (as shown explicitly in Fig. 6.1). The lengths of the anti-waveguiding and waveguiding sections are 4 and 2, respectively, the absolute value of the guiding parameter is $|A| = 4$ for both sections, the propagation parameter is $E = 2$, and $D = 0.8$ (recall D is defined in Eq. (6.3)).

Chapter 7

Stabilization of spatial solitons in bulk Kerr media with alternating nonlinearity

7.1   The model

The usual (2+1)-dimensional NLS equation governing the spatial evolution of light signals in bulk (3D) optical media with the $\chi^{(3)}$ (Kerr) nonlinearity cannot support stable spatial solitons in the form of cylindrical beams: if the nonlinearity is self-defocusing (negative), any beam spreads out, while in the case of the self-focusing (positive) non-linearity, a stationary-beam solution with a *critical value* of its norm (integral power) does exist, as was demonstrated by Chiao, Garmire, and Townes in 1964 (as a matter of fact, it was the first example of a soliton considered in nonlinear optics, and is frequently referred to as a *Townes soliton*, TS), but it is (weakly) unstable because of the possibility of the weak collapse in the 2D NLS equation [29, 159]. In work [30], it was shown, by means of direct simulations, that the beam can be partly stabilized if the nonlinearity coefficient in the 2D NLS equation is subjected to weak spatial modulation along the propagation direction, so that the beam's power (which is virtually constant, as radiative losses turn out to be negligible) effectively oscillates about the accordingly modulated critical value, being sometimes slightly above and sometimes slightly below it. As a result, it was observed that the beam could survive over a large propagation distance, although eventually is was destroyed by the instability.

A model of a strong NLM (nonlinearity management) for the bulk medium, with the periodic alternation of the sign of the local Kerr coefficient (the same as in the 1D NLM model in Eq. (4.11)), was proposed in paper [165]:

$$iu_z + \frac{1}{2}\nabla_\perp^2 u + \gamma(z)|u|^2 u = 0, \tag{7.1}$$

$$\gamma(z) = \begin{cases} \gamma_+, & \text{if } 0 < z < L_+, \\ \gamma_-, & \text{if } L_+ < z < L_+ + L_-, \end{cases} \qquad (7.2)$$

where the diffraction operator $\nabla_\perp^2$ acts on the transverse coordinates x and y. The main result of work [165] is that this model provides for *complete stabilization* of the (2+1)-dimensional solitons, as shown in detail below. The consideration of this model was not only interesting in its own right, but also served as a pattern for prediction of the stabilization of 2D BECs under the action of the time-periodic FRM, see the next chapter.

Axisymmetric spatial solitons, as solutions to Eq. (7.1), are sought for in the form

$$u(z, r, \theta) = \exp(iS\theta)U(z, r), \qquad (7.3)$$

where r and θ are the polar coordinates in the (x, y) plane, S is an integer vorticity ("spin"), and the function $U(z, r)$ obeys the equation

$$iU_z + \frac{1}{2}\left(U_{rr} + \frac{1}{r}U_r - \frac{S^2}{r^2}U\right) + \gamma(z)|U|^2 U = 0. \qquad (7.4)$$

7.2   Variational approximation

First of all, VA may be applied to Eq. (7.4). To this end, the following *ansatz* is adopted,

$$U = A(z)r^S \exp\left[ib(z)r^2 + i\phi(z)\right] \operatorname{sech}\left(\frac{r}{a(z)}\right), \qquad (7.5)$$

where A, b and a are the soliton's amplitude, chirp and width, and ϕ is the phase, cf. the 1D ansatz (2.6). An essential difference in the 2D case is the necessity to add the multipliers r^S, for the case of $S \neq 0$ (by definition, S is positive). Then, the following set of variational equations for the parameters of the ansatz (7.5) can be derived. First, due to the conservation of energy, there is a dynamical invariant

$$A^2 a^{2(S+1)} \equiv E, \qquad (7.6)$$

which makes it possible to eliminate the amplitude A from the equations. After that, the VA reduces to a second-order equation for $a(z)$,

$$\frac{d^2 a}{dz^2} = 2\left[\frac{I_2^{(S)}}{I_1^{(S)}} - \frac{I_4^{(S)}}{I_1^{(S)}}\gamma(z)\right]a^{-3}, \qquad (7.7)$$

the chirp being expressed as $b(z) = (2a)^{-1}da/dz$, cf. similar variational equations (2.12) and (2.11) for the 1D soliton. Numerical constants $I_{1,2,4}^{(S)}$ are integrals resulting from VA; for $S = 0$ (zero-spin beam), they are $I_{1,2,4}^{(0)} \approx (1.352, 0.398, 0.295)$.

In the NLM model with the periodic modulation (7.2), equation (7.7) can be solved inside each interval where γ is constant. The result is

$$\left(\frac{dV}{dz}\right)^2 + \Gamma = HV, \qquad (7.8)$$

where $V \equiv a^2$, H (which is actually the Hamiltonian of Eq. (7.7) with $\gamma = \text{const}$) is an arbitrary integration constant, and

$$\Gamma \equiv 8 \frac{I_2^{(S)} - I_4^{(S)} \gamma}{I_1^{(S)}}. \tag{7.9}$$

Within the interval $0 < z < L_+$, the parameter Γ keeps a constant value Γ_+, then it assumes another constant value Γ_- in the interval $L_+ < z < L_+ + L_-$, and this configuration repeats itself periodically. The formulas can be additionally rescaled to set $L_- \equiv 1$ and $\Gamma_- \equiv 1$, leaving one with two irreducible control parameters, $L_+ \equiv L$ and $\Gamma_+ \equiv \Gamma$ (note that the definition of Γ implies that, once it was set $\Gamma_- = 1$, then $\Gamma \equiv \Gamma_+$ may only take values smaller than 1, including negative values). Across each junction point where γ flips its sign, the values of V and dV/dz are related according to the physical conditions that the width and chirp of the pulse, as functions of z, must be continuous. As immediately follows from the above equations, this simply means that both V and dV/dz are continuous across the jump. Such boundary conditions are essentially easier to handle than their counterparts in the case of DM, where the continuity of the chirp, given by the expression $b = -2\beta(z)a(da/dz)$ (see Eq. (2.11)), and the jump of the GVD coefficient $\beta(z)$ at the junction point imposes a jump condition on the derivative da/dz. The simplification of the boundary conditions in the present NLM model makes it possible to obtain results in a completely analytical form, as shown below (which was impossible in the DM model).

Starting with arbitrary initial values $\tilde{V}_0$ and $\tilde{V}_0'$ of $V(z)$ and dV/dz at $z = 0$, one can derive a *map* that yields the values $\tilde{V}_0$ and $\tilde{V}_0'$ of the same variables at the end of the period, i.e., at $z = L_+ + L_- \equiv 1 + L$. Straightforward integration of Eq. (7.8) in the segments $L_\pm$, with regard to the continuity of V and dV/dz at the junction points, makes it possible to derive the map in an explicit although rather cumbersome form. Nevertheless, a *fixed point* (FP) of the map, which corresponds to the quasi-stationary propagation of the beam, can be found in a simple form:

$$V_0 = \pm \frac{L(\Gamma - 1)}{4\sqrt{L+1}\sqrt{-1 - L\Gamma}}, \quad V_0' = \mp \frac{\sqrt{-1 - L\Gamma}}{\sqrt{L+1}}. \tag{7.10}$$

This FP exists only for negative values of the coefficient (7.9), $\Gamma < -1/L$.

Calculating the *path-average* value of the nonlinearity coefficient, with regard to the normalizations adopted here,

$$\bar{\gamma} \equiv \frac{L_+\gamma_+ + L_-\gamma_-}{L_+ + L_-} = \frac{8I_2(L+1) - I_1\Gamma_-(L\Gamma + 1)}{8I_4(L+1)}, \tag{7.11}$$

it is easy to see that FPs (7.10) may only exist with *positive* $\bar{\gamma}$ (corresponding to self-focusing on average). Taking into account the above-mentioned necessary condition $\Gamma < -1/L$ and normalizations, Eq. (7.11) predicts the minimum value of the path-average nonlinearity coefficient at which the FPs exist: $(\bar{\gamma})_{\min} = I_2^{(0)}/I_4^{(0)} \approx 1.3453$. This result is quite natural, as it would be strange to expect the existence of quasi-stationary soliton beams in the case when the average nonlinearity is self-defocusing on average. On the other hand, it is relevant to recall that 1D stable solitons do exist in

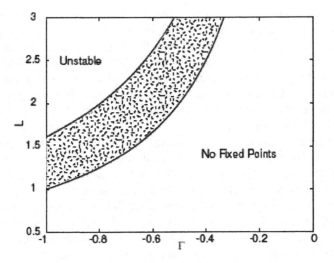

Figure 7.1: The existence and stability regions of the fixed points in the parameter plane (Γ, L) of the variational approximation for the model of the beam propagation in the layered bulk medium based on Eqs. (7.1) and (7.2). The fixed point is stable in the speckled area.

the DM model when the PAD coefficient is exactly zero or falls into a narrow interval of normal-dispersion values, see Fig. 2.5. This feature once again stresses the difference between the DM model and the present NLM one.

To investigate the stability of the FP, one should find eigenvalues μ of the map's Jacobian, $\partial \left(\tilde{V}_0, \tilde{V}_0' \right) / \partial \left(V_0, V_0' \right)$ (alias *multiplicators*). FP is stable if both eigenvalues satisfy the condition $|\mu| \leq 1$. The results of this analysis are summarized in Fig. 7.1. No FP exists beneath the curve $L = -1/\Gamma$. Above this line, FP is stable inside the speckled band. Outside the band, the FP exists but is unstable, according to the calculation of the multiplicators.

7.3   Numerical results

To test the VA-based analytical results against direct simulations, the underlying equation (7.1) was solved numerically, using the ansatz (7.5), with the parameters taken at the FP (7.10), as the initial configuration. A typical result is shown in Fig. 7.2: after

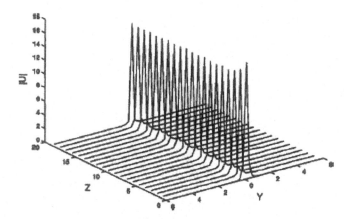

Figure 7.2: The evolution of the soliton's cross section in direct simulations of Eqs. (7.1), (7.2), as it propagates in z. As is seen, the initial configuration taken according to the variational ansatz (7.5) with $S = 0$ rapidly relaxes to a stable soliton beam. In this case, $L = 2.0$ and $\Gamma = -0.55$.

a short relaxation period, the initial beam reshapes into a nearly stationary stable one, which propagates with small residual oscillations.

On the other hand, simulations starting with the ansatz (7.5) that carries nonzero vorticity S show that, unlike the $S = 0$ beam, they all are unstable. In particular, the vortex beam with $S = 1$ splits into two stable fundamental solitons (ones corresponding to $S = 0$), which is a typical manifestation of the azimuthal instability of vortical solitons in media with "simple" nonlinearities (vortices may be stabilized in media with *competing nonlinearities*, such as self-focusing cubic and self-defocusing quintic [44], or quadratic and self-defocusing cubic [111]).

A 3D counterpart of the model (7.1) was investigated too. It is based on the equation

$$iu_z + \frac{1}{2}\left(\nabla_\perp^2 - \beta\frac{\partial^2}{\partial\tau^2}\right)u + \gamma(z)|u|^2u = 0, \tag{7.12}$$

where τ and β are, as usual, the reduced time and GVD coefficient, cf. Eq. (1.5), and $\gamma(z)$ is the same as in Eq. (7.2). To allow the existence of 3D solitons (STSs), β must be negative (corresponding to the anomalous dispersion). However, direct simulations demonstrate that stable 3D solitons are impossible in Eq. (7.12) [165].

Chapter 8

Stabilization of two-dimensional solitons in Bose-Einstein condensates under Feshbach-resonance management

The possibility to stabilize (2+1)-dimensional spatial solitons by means of the periodic alternation of the sign of the Kerr coefficient in a layered bulk material suggest a physically different but mathematically similar possibility to stabilize 2D soliton-like BEC configurations by means of the ac-FRM technique, which amount to making the nonlinearity coefficient in the corresponding 2D Gross-Pitaevskii equation (GPE) a sinusoidal function of time. This possibility was first independently explored in works [5] and [150]. Additional results, including the case of a 3D condensate strongly confined in one direction, which makes it nearly two-dimensional, were reported in paper [130], and a similar mechanism providing for stabilization of a two-component 2D soliton in a system of two nonlinearity coupled GPEs was investigated too [128]. Here, main results for this important problem will be presented, chiefly following paper [5]. It will also be shown that the ac-FRM technique (if acting alone) cannot stabilize 3D solitons.

8.1   The model and variational approximation

8.1.1   General consideration

The normalized GPE is taken in the usual form (cf. Eq. (5.5) in the 1D case),

$$iu_t + \frac{1}{2}\Delta u + [\lambda_0 + \lambda_1 \sin(\omega t)] |u|^2 u, \tag{8.1}$$

where Δ is the 2D or 3D Laplacian, λ_1 and ω being the ac-FRM amplitude and frequency. Further, using the scaling invariance of the equation, it is set $|\lambda_0| \equiv 1$, so that λ_0 is a sign-defining parameter, $\lambda_0 > 0$ and $\lambda_0 < 0$ corresponding to self-attraction and self-repulsion, respectively. The external potential is not included, with the intention to focus on a possibility to stabilize the condensate in the *trapless* case, relying solely on the FRM ac drive.

For the subsequent analysis, it is sufficient to consider solutions to Eq. (8.1) in the axially or spherically symmetric situation (in the 2D and 3D cases, respectively), assuming that the wave function u is a function of time and 2D or 3D radial variable r, without angular dependences (if the solution carries no vorticity, there is no danger of an isotropy-breaking instability, hence the angular dependences may be consistently ignored). The accordingly restricted equation (8.1) takes the form

$$i\frac{\partial u}{\partial t} = -\left(\frac{\partial^2}{\partial r^2} + \frac{D-1}{r}\frac{\partial}{\partial r}\right) u - [\lambda_0 + \lambda_1 \sin(\omega t)] |u|^2 u, \tag{8.2}$$

where $D = 2$ or 3 is the spatial dimension.

An analytical approach to Eq. (8.2) may be based on its variational representation, with the corresponding Lagrangian

$$L = \text{const} \int_0^\infty \mathcal{L}\{u, u^*, u_t, u_t^*\} \, r^{D-1} dr, \tag{8.3}$$

$$\mathcal{L} = \frac{i}{2}\left(\frac{\partial u}{\partial t}u^* - \frac{\partial u^*}{\partial t}u\right) - \left|\frac{\partial u}{\partial r}\right|^2 + \frac{1}{2}\lambda(t)|u|^4. \tag{8.4}$$

The respective variational *ansatz* for the wave function is based, as usual, on the Gaussian (cf. Eq. 7.5),

$$u_{\text{VA}}(r, t) = A(t)\exp\left(-\frac{r^2}{2a^2(t)} + \frac{1}{2}ib(t)\,r^2 + i\delta(t)\right), \tag{8.5}$$

where A, a, b and δ are, respectively, the amplitude, width, chirp and phase, which are assumed to be real functions of time.

8.1.2   The two-dimensional case

In the 2D case, the substitution of the ansatz (8.5) in Eqs. (8.3) and (8.4) and integration yield the effective Lagrangian,

$$L_{\text{eff}}^{(2D)} = \text{const} \cdot \left[-\frac{1}{2}a^4 A^2 \frac{db}{dt} - a^2 A^2 \frac{d\delta}{dt} - A^2 - a^4 A^2 b^2 + \frac{1}{4}\lambda(t)\,a^2 A^4\right]. \tag{8.6}$$

where $\lambda(t) \equiv \lambda_0 + \lambda_1 \sin(\omega t)$. The variational (Euler-Lagrange) equations following from this Lagrangian yield the conservation of the solution's norm (which is tantamount to the scaled number of atoms in the condensate),

$$\pi A^2 a^2 \equiv N = \text{const},\tag{8.7}$$

an expression for the chirp, $b = (2a)^{-1} da/dt$, and a closed-form evolution equation for the width:

$$\frac{d^2 a}{dt^2} = \frac{-\Lambda + \epsilon \sin(\omega t)}{a^3},\tag{8.8}$$

$$\Lambda \equiv 2\left(\lambda_0 \frac{N}{2\pi} - 2\right), \epsilon \equiv -\frac{\lambda_1 N}{\pi}.\tag{8.9}$$

Actually, these equations are the same as ones presented (also for the 2D case) in the previous chapter, see Eqs. (7.6) and (7.7), differing only in the form of the modulation functions $\lambda(t)$ and $\gamma(z)$.

In the absence of the time-periodic (ac) modulation, $\epsilon = 0$, Eq. (8.8) conserves the Hamiltonian,

$$H_{2D} = \frac{1}{2}\left[\left(\frac{da}{dt}\right)^2 - \frac{\Lambda}{a^2}\right].\tag{8.10}$$

Obviously, $H_{2D} \to -\infty$ as $a \to 0$, if $\Lambda > 0$, and $H_{2D} \to +\infty$ as $a \to 0$, if $\Lambda < 0$. This means that, in the absence of the ac drive, the 2D pulse is expected to collapse if $\Lambda > 0$, and to spread out if $\Lambda < 0$. The case $\Lambda = 0$ corresponds to the critical norm which is the separatrix between the collapsing and decaying solutions. The critical norm corresponds to the solution in the form of the above-mentioned Townes soliton (TS). Note that a numerically exact value of the critical norm is $N = 1.862$ (in the present notation, and setting, as said above, $\lambda_0 = +1$) [29, 159], while the variational equation (8.9) yields, as the analytical approximation for it, $N = 2$.

If the ac component of the nonlinear coefficient oscillates at a high frequency, Eq. (8.8) can be treated analytically by means of the averaging method (which may be compared to the derivation of Eq. (5.15) in the 1D model with rapid modulation of the trap's strength). To this end, one sets $a(t) = \bar{a} + \delta a$, with $|\delta a| << |\bar{a}|$, where $\bar{a}$ varies on a slow time scale and δa is a rapidly varying function with a zero mean value. After straightforward manipulations, the following equations for the slow and rapid variables can be derived:

$$\frac{d^2}{dt^2}\bar{a} = -\Lambda(\bar{a}^{-3} + 6\bar{a}^{-5}\langle(\delta a)^2\rangle) - 3\epsilon\langle\delta a \sin(\omega t)\rangle\bar{a}^{-4},\tag{8.11}$$

$$\frac{d^2}{dt^2}\delta a = 3\delta a\Lambda\bar{a}^{-4} + \epsilon \sin(\omega t)\bar{a}^{-3}.\tag{8.12}$$

where $\langle...\rangle$ stands for averaging over the period $2\pi/\omega$. A solution to the linear equation (8.12) is straightforward,

$$\delta a(t) = -\frac{\epsilon \sin(\omega t)}{\bar{a}^3(\omega^2 + 3\bar{a}^{-4}\Lambda)}.\tag{8.13}$$

The substitution of this into Eq. (8.11) and averaging yield the final evolution equation for the slow variable,

$$\frac{d^2}{dt^2}\bar{a} = \bar{a}^{-3}\left[-\Lambda - \frac{3\Lambda\epsilon^2}{(\omega^2\bar{a}^4 + 3\Lambda)^2} + \frac{3}{2}\frac{\epsilon^2}{\omega^2\bar{a}^4 + 3\Lambda}\right]. \tag{8.14}$$

To understand whether collapse is enforced or inhibited by the ac modulation of the nonlinearity, one may consider Eq. (8.14) in the limit $\bar{a} \to 0$, when it reduces to

$$\frac{d^2}{dt^2}\bar{a} = (-\Lambda + \frac{\epsilon^2}{6\Lambda})\bar{a}^{-3}. \tag{8.15}$$

It immediately follows from Eq. (8.15) that, if the amplitude of the high-frequency ac drive is large enough, $\epsilon^2 > 6\Lambda^2$, the behavior in the limit of small $\bar{a}$ is exactly opposite to that which would be expected in the presence of the dc component only: in the case of $\Lambda > 0$, bounce from the center should occur instead of the collapse, and vice versa in the case of $\Lambda < 0$.

On the other hand, in the limit of large $\bar{a}$, Eq. (8.14) takes the asymptotic form $d^2\bar{a}/dt^2 = -\Lambda/\bar{a}^3$, which shows that the condensate remains self-confined in the case of $\Lambda > 0$ (the negative acceleration $d^2\bar{a}/dt^2$ implies that the variable $\bar{a}$ is pulled back from large to smaller values). Thus, these asymptotic results suggest that Eq. (8.14) gives rise to a *stable behavior* of the condensate, with both the collapse and decay being ruled out if

$$\epsilon > \sqrt{6}\Lambda > 0. \tag{8.16}$$

In other words, conditions (8.16) ensures that the right-hand side of Eq. (8.14) is positive for small $\bar{a}$ and negative for large $\bar{a}$, which implies that Eq. (8.14) must give rise to a stable FP (fixed point). Indeed, when the conditions (8.16) hold, the right-hand side of Eq. (8.14) vanishes at exactly one FP,

$$\omega^2\bar{a}^4 = \frac{3\epsilon^2}{4\Lambda} + \sqrt{3\left(\frac{3\epsilon^4}{16\Lambda^2} - 1\right)} - 3\Lambda, \tag{8.17}$$

which can be easily checked to be stable, within the framework of Eq. (8.14), through the calculation of an eigenfrequency of small oscillations around it.

Direct numerical simulations of ODE (8.8) produce results (not shown here) which are in exact correspondence with those predicted by the averaging method, i.e., a stable state with $a(t)$ performing small oscillations around the point (8.17).

8.1.3   Variational approximation in the three-dimensional case

The calculation of the effective Lagrangian (8.3), (8.4) with the ansatz (8.5) in the 3D case yields

$$L_{\text{eff}}^{(3D)} = \frac{1}{2}\pi^{3/2}A^2a^3\left(-\frac{3}{2}\frac{db}{dt}a^2 - 2\frac{d\delta}{dt} + \frac{1}{2\sqrt{2}}\lambda(t)A^2 - \frac{3}{a^2} - 3b^2a^2\right), \tag{8.18}$$

cf. Eq. (8.6). The Euler-Lagrange equations applied to this Lagrangian again yield the norm conservation,

$$\pi^{3/2} A^2 a^3 \equiv N = \text{const} \tag{8.19}$$

(cf. the 2D counterpart (8.7)), the same expression for the chirp as in the 2D case, $b = (2a)^{-1} da/dt$, and the final evolution equation for the width of the condensate,

$$\frac{d^2 a}{dt^2} = \frac{4}{a^3} + \frac{-\Lambda + \epsilon \sin(\omega t)}{a^4}, \tag{8.20}$$

where the adopted definitions are $\Lambda \equiv \lambda_0 N/\sqrt{2\pi^3}$ and $\epsilon \equiv -\lambda_1 N/\sqrt{2\pi^3}$, cf. Eq. (8.9). Note the difference of Eq. (8.20) from its 2D counterpart (8.14).

In the absence of the ac drive, $\epsilon = 0$, Eq. (8.20) conserves the Hamiltonian

$$H_{3D} = \frac{1}{2} \left(\frac{da}{dt} \right)^2 + \frac{2}{a^2} - \frac{\Lambda}{3a^3}. \tag{8.21}$$

Obviously, $H_{3D} \to -\infty$ as $a \to 0$, if $\Lambda > 0$, and $H_{3D} \to +\infty$ if $\Lambda < 0$, hence one will have collapse or decay of the pulse, respectively, in these two cases.

ODE (8.20) was solved numerically (without averaging), to check if (within the framework of the VA) there is a region in the parameter space where the condensate, that would decay under the action of the repulsive dc nonlinearity ($\Lambda < 0$), may be stabilized by the ac FRM drive in the 3D case. Figure 8.1 shows the dynamical behavior of solutions to Eq. (8.20) in terms of the Poincaré section in the plane of the variables $(a, a' \equiv da/dt)$, for $\Lambda = -1, \epsilon = 100, \omega = 10^4 \pi$, and initial conditions $a(t = 0) = 0.3, 0.2,$ or 0.13 and $a'(t = 0) = 0$. It is seen from the figure that, in all these cases, the solution remains bounded and the condensate does not collapse or decay, its width performing quasi-periodic oscillations.

Systematic simulation of Eq. (8.20) demonstrate that, as well as in the examples shown in Fig. 8.1, the frequency and amplitude of the ac drive need to be large to secure the stability, which suggests to apply the averaging procedure to this case too (similar to how it was done above for the 2D case). The stability is predicted by the simulations only for $\Lambda < 0$, i.e., for the repulsive dc (constant) component of the nonlinearity. In the opposite case, of $\Lambda > 0$ (attractive dc nonlinearity), the VA predict only collapse, i.e., the situation is exactly opposite to that in two dimensions, where stability was predicted solely for $\Lambda > 0$, see Eq. (8.16).

The averaging procedure goes through the calculation of the rapidly oscillating correction $\delta a(t)$,

$$\delta a = -\frac{\epsilon \sin(\omega t) \bar{a}}{\omega^2 \bar{a}^5 - 12\bar{a} + 4\Lambda}, \tag{8.22}$$

cf. Eq. (8.13) in the 2D case. The final averaged equation for the slow variable $\bar{a}(t)$ is (cf. Eq. (8.14))

$$\frac{d^2 \bar{a}}{dt^2} = \bar{a}^{-4} \left[4\bar{a} - \Lambda + \frac{2\epsilon^2}{\omega^2 \bar{a}^5 - 12\bar{a} + 4\Lambda} + \epsilon^2 \frac{6\bar{a} - 5\Lambda}{(\omega^2 \bar{a}^5 - 12\bar{a} + 4\Lambda)^2} \right]. \tag{8.23}$$

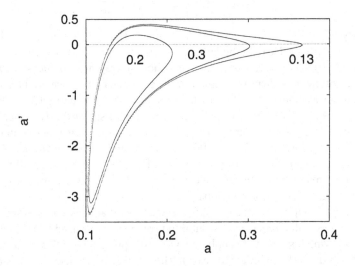

Figure 8.1: The Poincaré section in the plane $(a, a' \equiv da/dt)$, generated by the numerical solution of the variational equation (8.20) in the three-dimensional model of the ac-FRM-driven Bose-Einstein condensate for $\Lambda = 1$, $\epsilon = 100$, $\omega = 10^4\pi$ and different initial conditions. The full model is based on the Gross-Pitaevskii equation (8.1).

In the limit of $\bar{a} \to 0$, Eq. (8.23) takes the form

$$\frac{d^2\bar{a}}{dt^2} = \left(-\Lambda + \frac{3\epsilon^2}{16\Lambda}\right)\bar{a}^{-4}, \tag{8.24}$$

cf. Eq. (8.15). Equation (8.24) predicts one feature of the 3D model correctly, *viz.*, in the case of $\Lambda < 0$ and with a sufficiently large amplitude of the ac component, $\epsilon > \left(4/\sqrt{3}\right)|\Lambda|$, collapse takes place instead of the decay. However, other results following from the averaged equation (8.23) are *wrong*, as compared to those following from direct simulations of the full variational equation (8.20), some of which are displayed in Fig. 8.1. In particular, detailed analysis of the right-hand side of Eq. (8.23) shows that it *does not* predict a stable FP for $\Lambda < 0$, and *does predict* it for $\Lambda > 0$, exactly opposite to what was revealed by direct simulations of the underlying ODE (8.20). This failure of the averaging approach (in stark contrast with the 2D case) may be explained by the existence of singular points in Eqs. (8.22) and (8.23) (for both $\Lambda > 0$ and $\Lambda < 0$), at which the denominator $\omega^2 \bar{a}^5 - 12\bar{a} + 4\Lambda$ in these equations vanishes. Note that, in the 2D case with $\Lambda > 0$, for which the stable state in the region (8.16) was predicted above, the corresponding equation (8.14) did not have such singularities.

8.2 Averaging of the Gross-Pitaevskii equation and Hamiltonian

In the case of the high-frequency FRM modulation, there is a possibility to apply the averaging method directly to the GPE (8.2), without the resort to the VA. To this purpose, the solution to the PDE is looked for as an expansion in powers of $1/\omega$,

$$u(r, t) = A_0(r, t) + \omega^{-1}A_1(r, t) + \omega^{-2}A_2(r, t) + ..., \tag{8.25}$$

with $\langle A_{1,2,...}\rangle = 0$ (the symbol $\langle...\rangle$ stands for the average over the period of the rapid modulation). The normalization $\lambda_0 = +1$ is adopted here, as it is expected that it should provide for stability in the 2D case. The final result is an effective equation for the main (slowly varying) part of the wave function in the expansion (8.25) derived at the order ω^{-2} [5]:

$$i\frac{\partial A_0}{\partial t} + \Delta A_0 + |A_0|^2 A_0 + \frac{1}{2}\lambda_1^2\left(\frac{\epsilon}{\omega}\right)^2 [|A_0|^6 A_0 - \tag{8.26}$$

$$3|A_0|^4 \Delta A_0 + 2|A_o|^2 \Delta(|A_0|^2 A_0) + A^2 \Delta\left(|A_0|^2 A_0^*\right)] = 0, \tag{8.27}$$

where ϵ is the same amplitude as in Eq. (8.9). Note that Eq. (8.27) is valid in the 2D and 3D cases alike. Cumbersome analysis of this equation, performed in work [5] demonstrates that the collapse is indeed arrested in the 2D version of the equation, essentially the same way as it was predicted above by the VA.

To understand the nature of the collapse arrest by the high-frequency FRM drive, it is actually more instructive to insert the expansion (8.25) and perform the subsequent averaging not in equation (8.2) itself, but rather in the Hamiltonian,

$$H = C \int_0^\infty \left(-\left| \frac{\partial u}{\partial r} \right|^2 + \frac{1}{2} \lambda(t) |u|^4 \right) dV, \qquad (8.28)$$

where dV is the infinitesimal volume in the 2D or 3D case, and a constant C is positive. The resulting averaged Hamiltonian, expressed in terms of the slowly varying main part $A_0(r, t)$ of the wave function, is [5]

$$\bar{H} = \int dV \left[|\nabla A_0|^2 - \frac{1}{2} |A_0|^4 + \frac{\lambda_1^2}{2} \left(\frac{\epsilon}{\omega} \right)^2 \left(|\nabla(|A_0|^2 A_0)|^2 - 3|A_0|^8 \right) \right], \qquad (8.29)$$

where λ_1 is the same coefficient as in Eq. (8.1).

A possibility to arrest the collapse can be explored using the effective Hamiltonian (8.29). To this end, one may follow the pattern of the usual *virial estimates* [29, 159]. Thus, one notes that, if a given field configuration has compressed itself to a spot with a small size ρ and large amplitude $\aleph$, the conservation of the norm N (in the first approximation, the norm conservation remains valid for the field A, as follows form the expansion (8.25)) yields a relation

$$\aleph^2 \rho^D \sim N \qquad (8.30)$$

(recall D is the dimension of space). On the other hand, estimates of the same type, applied to the strongest collapse-driving and collapse-arresting terms, H_- and H_+, in the averaged Hamiltonian (which are the fourth and second terms, respectively, in expression (8.29)) yield

$$H_- \sim -\left(\frac{\epsilon}{\omega} \right)^2 \aleph^8 \rho^D, \quad H_+ \sim \left(\frac{\epsilon}{\omega} \right)^2 \aleph^6 \rho^{D-2}. \qquad (8.31)$$

Eliminating the amplitude from Eqs. (8.31) by means of the relation (8.30), one concludes that, in the case of the catastrophic self-compression of the field in the 2D space, $\rho \to 0$, both terms H_- and H_+ assume a common asymptotic form, $\sim \rho^{-6}$, hence the collapse may be stopped, depending on details of the initial configuration (the initial state determines a ratio between coefficients in front of ρ^{-6} in the two asymptotic expressions). On the contrary to this, in the 3D case the collapse-driving term H_- diverges as ρ^{-9}, while the collapse-arresting one has the asymptotic form $\sim \rho^{-8}$ (for $\rho \to 0$), hence in this case the collapse *cannot* be prevented.

8.3   Direct numerical results

The existence of stable 2D self-confined soliton-like oscillating condensate states, predicted above by means of analytical approximations in the region (8.16), when the dc part of the nonlinearity corresponds to attraction in the BEC, was checked against direct simulations of the 2D equation (8.2), i.e., one with $D = 2$ [5]. In was quite easy to confirm this prediction. In the case of $\lambda_0 = -1$ in Eq. (8.2), i.e., when the dc component of the nonlinearity corresponds to self-repulsion, direct simulations always

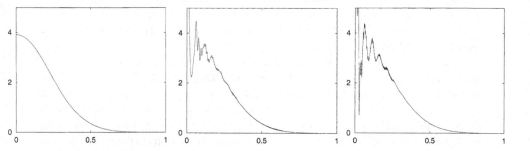

Figure 8.2: Time evolution of the condensate's shape $|u(r)|^2$, as obtained from direct simulations of the radial Gross-Pitaevskii equation (8.2) with strong and fast ac-FRM modulation ($\omega = 10^4\pi$, $\epsilon = 90$). From left to right, the panels pertain to the moments of time $t = 0.007, 0.01$ and 0.015.

show decay (spreading out) of the 2D condensate, which also agrees with the above predictions.

Additionally, direct simulations show that, unlike the fundamental solitons, their vortical counterparts cannot be stabilized by the FRM-induced periodic modulation of the nonlinearity. On the other hand, in a recent work [129] it was demonstrated that a two-component (vectorial) generalization of the present model may support stable existence, in the course of very long time, of compound solitons in which one component is arranged as a (partly incoherent) fundamental soliton, while the other one carries vorticity.

In the 3D case, direct simulations are still more necessary, in view of a rather controversial character of the variational results for this case, as explained above. With $\Lambda < 0$ (the repulsive dc nonlinearity), and a sufficiently large amplitude of the ac FRM drive, the simulations show *temporary stabilization* of the condensate, roughly the same way as predicted above by the solution of the variational equation (8.20). However, the stabilization *is not permanent*: the condensate begins to develop small-amplitude short-scale modulations around its center, and after $\simeq 50$ periods of the ac modulation it collapses. An example of this behavior is displayed in Fig. 8.2, for $N = 1$, $\Lambda = -1$ and $\omega = 10^4\pi$.

Results displayed in Fig. 8.2 are typical for the 3D case with $\Lambda < 0$. The eventual collapse which takes place in this case is a nontrivial feature, as it occurs despite the fact that the dc part of the nonlinearity drives the condensate towards spreading out. In the case of $\Lambda > 0$, simulations show that the ac-FRM drive is never able to prevent the collapse. These general conclusions comply with the analysis developed above on the basis of the averaged Hamiltonian (8.29), which showed that the collapse cannot be arrested in the 3D case, provided that the amplitude of the ac drive is large enough. Besides that, this eventual result is also in accordance with simulations of the model of the bulk optical medium with the periodically alternating sign of the Kerr coefficient, based on Eqs. (7.1) and (7.2): as mentioned in the previous chapter, simulations never produced stable 3D solitons (STSs) in this model.

Lastly, it is relevant to mention paper [14], where opposite conclusions were reported: that the ac-FRM modulation can stabilize 3D solitons, and can stabilize 2D vortical solitons too. The latter result was obtained in simulations of the GPE restricted to the temporal and radial variables only, similar to Eq. (8.2), while, as is known, the instability of soliton vortices is induced by azimuthal perturbations which break their axial symmetry [44, 111]. As concerns the stability of 3D FRM-driven solitons reported in work [14], it seems plausible that what was observed in that work is an unstable soliton which is made to seem stable if its shape is found with very high accuracy, and perturbations are kept extremely small in the simulations. All the other works on this topic [5, 150, 130, 166] concluded that 3D solitons in the ac-FRM-driven model are unstable if no additional stabilizing element is present. In section 10.2, it will be shown that a combination of the ac-FRM drive and quasi-one-dimensional optical lattice (OL) is sufficient for true stabilization of 3D solitons in the corresponding GPE.

Chapter 9

Multidimensional dispersion management

9.1 Models

As it was explained in the Introduction, up to date neither 3D spatiotemporal solitons (STS) in a bulk optical medium, nor their 2D counterparts in planar waveguides have been observed in the experiment. For this reason, theoretical analysis of new settings that may suggest possibilities to create optical STS in the experiment remains a highly relevant topic.

In paper [122], a new scheme capable to support stable STS in the case of the ordinary Kerr ($\chi^{(3)}$) nonlinearity in a 2D medium (planar waveguide) was proposed. It relies on propagation of an optical beam across a layered structure that does not affect the nonlinearity (i.e., the Kerr coefficient is positive and constant everywhere, on the contrary to models with the nonlinearity management considered in previous chapters), but rather provides for periodic reversal of the sign of the local GVD coefficient. Thus, it is a model of the DM (dispersion management) in a nonlinear planar waveguide. The same work [122] has concluded that the DM itself, without additional ingredients, *cannot* stabilized 3D solitons in a bulk medium with a similar layered structure. In the 2D case, however, not only the ordinary stable single-peaked (fundamental) solitons, but also very robust double-peaked localized oscillatory states were found. This chapter summarizes basic results concerning the stabilization of the STSs by means of the DM in two dimensions.

The model is a straightforward variant of the NLS equation, quite similar to ones considered in previous chapters (cf., e.g., Eq. (7.12)):

$$iu_z + \frac{1}{2}\left(u_{xx} + D(z)u_{\tau\tau}\right) + |u|^2 u = 0. \tag{9.1}$$

Here z is the propagation distance, x is the transverse coordinate in the planar waveguide (in the bulk medium, u_{xx} is replaced by $u_{xx} + u_{yy}$, where y is the another transverse coordinate), τ is the same reduced time as in Eq. (1.5), and the local GVD coefficients is denoted here as $-D$, instead of β. It is subject to the same DM modulation as in Eq. (1.49),

$$D(z) = \begin{cases} D_+ > 0, \ 0 < z < L_+, \\ D_- < 0, \ L_+ < z < L_+ + L_- \equiv L, \end{cases} \quad (9.2)$$

which repeats itself periodically with the period L (in terms of D, the anomalous and normal GVD correspond, respectively, to $D > 0$ and $D < 0$). Note that Eq. (9.1)) has a manifest property of the Galilean invariance: if $u_0(x, z, \tau)$ is a solution, a two-parametric family of "boosted" (moving) solutions can be generated from it as follows:

$$u(x, z, \tau) = \exp\left[i\left(qx - \omega\tau - \frac{1}{2}q^2 z - \frac{1}{2}\omega^2 \int D(z)dz\right)\right] \times$$
$$u_0\left(x - qz, z, \tau + \omega \int D(z)dz\right), \quad (9.3)$$

where q and ω are two arbitrary real parameters ("Galilean boosts"), which is to be compared with the similar invariance of the 1D NLS equation, see Eq. (1.6).

To cast the model into a normalized form, one can set, by means of obvious rescalings, $D_+ \equiv 1$, $L \equiv 2$. The ratio L_-/L_+ remains an irreducible parameter, but it is well known that, in the usual DM model for optical fibers, the results do not depend on this ratio, nor separately on the soliton's temporal width T_{FWHM}, but rather on the DM strength, $S \equiv (D_+L_+ + |D_-|L_-)/T_{\text{FWHM}}^2$, see Eq. (2.25). Therefore, it will be fixed here that $L_+ = L_- = 1$. Then, in addition to S, the only remaining free parameter of the model is the PAD (path-average dispersion), which takes the form

$$\overline{D} \equiv \frac{D_+L_+ + D_-L_-}{L} = \frac{1}{2}(1 + D_-), \quad (9.4)$$

with regard to the normalizations $D_+ = 1$, $L_\pm = 1$. The remaining parameter D_- can be expressed in terms of $\overline{D}$: as it follows from Eq. (9.4), $D_- = 2\overline{D} - 1$.

It is relevant to mention that this 2D model is somewhat similar to another one, that was recently introduced in work [2]; it differs by sinusoidal modulation of $D(z)$, instead of the piece-wise constant DM map adopted in Eq. (9.2), and, most importantly, by the fact that (in the present notation) it has the same modulated coefficient multiplying both the GVD term $u_{\tau\tau}$ and the diffraction one, u_{xx}:

$$iu_z + \frac{1}{2}[D_0 + D_1 \sin(kz)](u_{xx} + u_{\tau\tau}) + |u|^2 u = 0, \quad (9.5)$$

with $D_0 > 0$. In fact, this model was motivated by a continuum limit of some discrete systems; as concerns the context of nonlinear optics, it would be difficult to introduce the periodic reversal of the sign of the transverse diffraction (the coefficient in front of u_{xx}). There is a great difference between Eq. (9.1), which is strongly anisotropic in the plane (x, τ), and the isotropic equation (9.5).

9.2 Variational approximation

The VA (variational approximation) is a natural tool for the analysis of solitons in Eq. (9.1). To apply it, a straightforward Gaussian ansatz is adopted (cf. *ansätze* (2.6) and

(8.5)),

$$u = A(z) \exp\left\{ i\phi(z) - \frac{1}{2}\left[\frac{x^2}{W^2(z)} + \frac{\tau^2}{T^2(z)} \right] + \frac{i}{2}\left[b(z)\,x^2 + \beta(z)\,\tau^2 \right] \right\}, \quad (9.6)$$

where A and ϕ are the amplitude and phase of the soliton, W and T are its transverse and temporal widths (the latter is related to the above-mentioned full-width-at-half-maximum as follows: $T_{\mathrm{FWHM}} = 2\sqrt{\ln 2}\,T$), and b and β are the spatial and temporal chirps. The Lagrangian from which the 2D version of Eq. (9.2) is derived is

$$L = \frac{1}{2}\int_{-\infty}^{+\infty}\left[i\left(u_z u^* - u_z^* u\right) - |u_x|^2 - D(z)\,|u_\tau|^2 + |u|^4 \right]\,dx\,d\tau. \quad (9.7)$$

Substitution of the ansatz (9.6) into the Lagrangian yields an *effective Lagrangian*,

$$\frac{4}{\pi}L_{\mathrm{eff}} = A^2 W T \left[4\phi' - b'W^2 - \beta'T^2 - W^{-2} - DT^{-2} \right.$$
$$\left. + A^2 - b^2 W^2 - D(z)\beta^2 T^2 \right], \quad (9.8)$$

where the prime stands for d/dz.

The variational equation $\delta L/\delta\phi = 0$, applied to the expression (9.8), yields, as always, the energy-conservation relation, $dE/dz = 0$, where

$$E \equiv A^2 W T. \quad (9.9)$$

Equation (9.9) is used to eliminate A^2 from subsequent equations. Then, the term $\sim \phi'$ in the Lagrangian may be dropped, and it takes the form

$$\frac{4L_{\mathrm{eff}}}{\pi E} = -b'W^2 - \beta'T^2 - \frac{1}{W^2} - \frac{D(z)}{T^2} + \frac{E}{WT} - b^2 W^2 - D(z)\beta^2 T^2. \quad (9.10)$$

Varying the latter expression with respect to W, T and b, β yields the following Euler-Lagrange equations:

$$b = \frac{W'}{W}, \quad \beta = \frac{T'}{D(z)T}, \quad (9.11)$$

$$W'' = \frac{1}{W^3} - \frac{E}{2W^2 T}, \quad (9.12)$$

$$T'' - \frac{D'}{D}T' = \frac{D^2}{T^3} - \frac{DE}{2WT^2}. \quad (9.13)$$

Note that, as is known from the variational approach to the description of collapse in the 2D NLS equation [48], in the case of $\beta = \mathrm{const} < 0$, fixed-point (FP) solutions to the VA equations (9.12) and (9.13) are degenerate: the FP exists at a single value of the energy, $E = 2\sqrt{D}$, and, at this special value of E, there is a family of FPs with $T = \sqrt{D}W$ (W is arbitrary). These results exactly correspond to the existence of the TS (Townes soliton) in the isotropic 2D NLS equation. The family of the TS solutions are found at a single value of the energy, but with arbitrary width. Within the framework of Eqs. (9.12) and (9.13), all the FPs corresponding to $D = \text{const} > 0$ are weakly unstable (against small perturbations that grow with z linearly, rather than exponentially, which corresponds to the fact that the collapse is weak in the 2D case, on the contrary to strong collapse in three dimensions).

In the case of the piece-wise constant modulation corresponding to Eq. (9.2), the variables W, W', T and β at junctions between the segments with $D = D_\pm$ must be continuous. As it follows from Eq. (9.11), the continuity of the temporal chirp $\beta(z)$ implies a jump of T' when passing from D_- to D_+, or vice versa:

$$(T')_{D=D_+} = \frac{D_+}{D_-} (T')_{D=D_-}. \qquad (9.14)$$

9.3   Numerical results

Both the variational equations (9.12), (9.13) and the underlying GPE (9.1) were simulated numerically. In the latter case, the initial state was taken as per the ansatz (9.6), with zero spatial chirp (obviously, a point at which it vanishes can always be found, so this choice does not imply any special restriction), while the initial temporal chirp, β_0, was included:

$$u_0 = A_0 \exp\left[-\frac{1}{2}\left(\frac{x}{W_0}\right)^2 + \left(\frac{\tau}{T_0}\right)^2 - i\beta_0\,\tau^2\right]. \qquad (9.15)$$

Completely stable periodically oscillating solitons were easily found in direct PDE simulations, closely following the prediction provided by the VA. As a typical example, Fig. 9.1 shows a sequence of the soliton's snapshots through one (40th) cycle of the evolution. This picture remains identically the same, for instance, at the 80th cycle, attesting to the true stability of the solitons in the 2D DM model. No leakage (net radiation loss) from the established soliton was observed, up to the accuracy of numerical simulations. This implies that a small amount of radiation, emitted from the pulse when it passes the normal-dispersion slice, is absorbed back into it in the slice with anomalous dispersion.

The shape of the pulse in Fig. 9.1 remains very close to a Gaussian, which explains why the VA provides for good accuracy in this case. The evolution of the temporal width $T(z)$ for the same parameters, as predicted by the VA, is displayed in Fig. 9.2. On the contrary to $T(z)$, the spatial width $W(z)$ remains nearly constant, suggesting that the stable 2D soliton may be realized, in loose terms, as a "product" of the temporal DM soliton in the τ-direction, and ordinary spatial soliton localized in x (a factorized ansatz of this type was proposed for the consideration of multidimensional optical solitons in work [97]). Comparison shows that the PDE numerical solution indeed agrees well with the VA prediction.

However, in some other cases direct simulations of Eq. (9.1) reveal periodic evolution in a drastically different form: the initial pulse splits into two subpulses, which do not fully separate, but rather form an oscillatory bound state, examples of which are

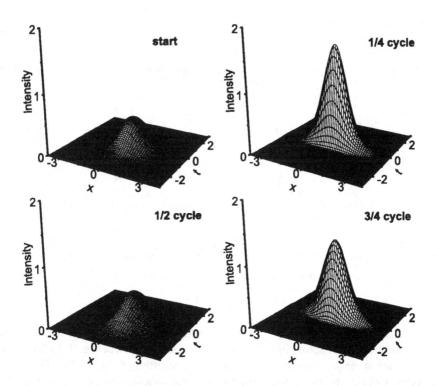

Figure 9.1: Evolution of the intensity distribution in a stable 2D soliton through a cycle of its propagation in the dispersion-managed nonlinear planar waveguide, described by Eqs. (9.1) and (9.2), with $D_+ = -D_- = 1$, $L_+ = L_- = 1$. Parameters of the initial pulse (9.15) are $T_0 = 1.35$, $W_0 = 1.35$, $E = 1$, and $\beta_0 = -1.85$. Snapshots are taken at points corresponding to the start, 1/4, 1/2 and 3/4 of the cycle.

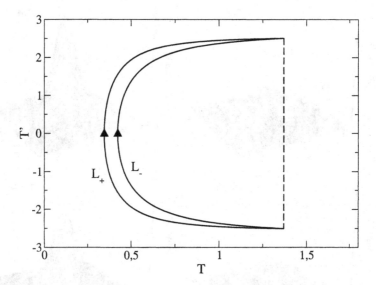

Figure 9.2: A cycle of the soliton's evolution in the plane of $(T, T' \equiv dT/dz)$, according to the variational approximation, in the same case as shown in Fig. (9.1). The jump in T' occurs at the junction between L_+ and L_-, according to Eq. (9.14). Unlike the temporal width T, its spatial counterpart W remains almost constant within the cycle.

shown in Figs. 9.3. In this case the VA still predicts a stable soliton in the form of a single Gaussian. Oscillatory bound states of two subpulses are also found in some cases when the VA predicts that the Gaussian-shaped soliton cannot self-trap. The bound state demonstrates qualitatively similar behavior in all the cases when it is found. First, an initial Gaussian with zero temporal chirp splits into two subpulses with chirps of opposite signs. Then, in the regime of established oscillations, the subpulses approach each other and nearly merge while passing the layer with $D = D_+$, and they separate again in the layer with $D = D_-$.

The VA makes wrong predictions in cases when the pulse transforms itself into the bound state of two subpulses, as the simple Gaussian ansatz (9.6) obviously cannot describe such a configuration. It is relevant to mention that the splitting of an initial Gaussian is one of possible generic outcomes of the evolution in 1D models of the DM type, see Fig. 2.3. However, a cardinal difference is that no stable oscillatory bound states resulting from the splitting was found in 1D models. In this connection, it may be relevant to mention that a drastic difference of the splitting of 1D (temporal) pulses and their spatiotemporal counterparts was observed in a recent experimental work [119], which was dealing with the propagation of ultrashort spatiotemporal pulses in water: while the pulse undergoes on-axis splitting and recombination, its spatially integrated temporal profile remains unsplit.

The findings are summarized in stability diagrams for the 2D solitons displayed in Fig. 9.4. The diagrams were generated on the basis of simulations of the variational equations (9.12) and (9.13), which were verified by direct simulations of the PDE (9.1) at sampling points indicated in the diagrams by digits. At points 1, 2, 3, 6, 9, and 10 the behavior predicted by the VA is confirmed by the simulations. At points 7 and 8, the periodic split-pulse evolution (bound states) is observed, of the type shown in Figs. (9.3). Note that this behavior, which may be interpreted, in loose terms, as intermediate between the stability and decay of a single-peaked soliton, is indeed observed close to VA-predicted borders between stable and decaying solitons.

At point 4, which is close to the VA-predicted border between decay and collapse, direct simulations initially demonstrate strong emission of radiation and broadening of the pulse, which eventually cease, being changed by seemingly chaotic oscillations of the localized pulse without any tangible energy loss. At point 5, essentially the same chaotic regime sets in, which is preceded, however, by self-compression of the initial pulse, rather than by its broadening. Lastly, at point 11, a strong transient emission of radiation is observed, like at point 4, but the pulse keeps its Gaussian shape, and regular periodic oscillations of the soliton finally set in. It may happen that, in the course of an extremely long evolution, chaotically oscillating solitons (observed at points 4 and 5) gradually relax towards a periodically oscillating soliton.

Finally, the observation that the stability regions (unshaded in Fig. 9.4) tend to be centered around a particular value of the energy (especially in Fig. 9.4(b)) has a simple explanation: this value is the one which corresponds to the TS (Townes soliton) in the free space, and, in a crude approximation, the DM, as well as other means of stabilization of the 2D solitons (such as 2D and quasi-1D lattices [24, 26]) act to stabilize the pre-existing weakly unstable TS.

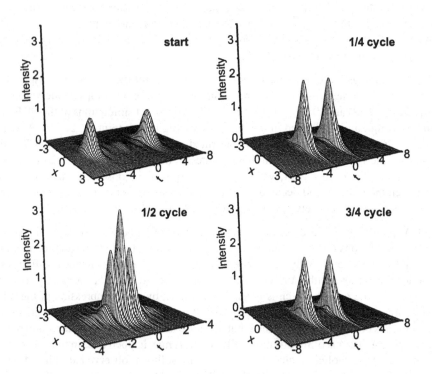

Figure 9.3: The same as in Fig. 9.2, but for different parameters of the input pulse: $T_0 = 1$, $W_0 = 1$, $E = 2$, and $\beta_0 = 0$. In this case, although the variational approximation predicts a stable single-peaked solution, the pulse splits up and gives rise to a stable oscillatory bound state of two subpulses.

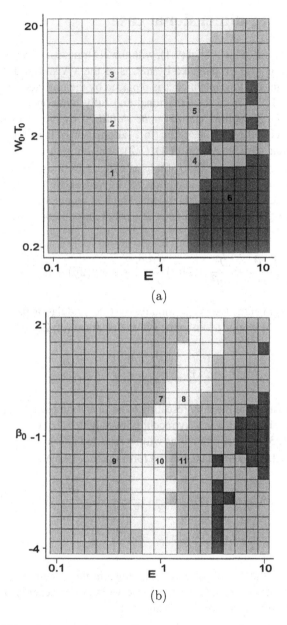

Figure 9.4: Stability diagrams for the solitons in the model of the dispersion-managed nonlinear planar waveguide: (a) in the plane (E, W_0) of the energy and width of the initial pulse, with $W_0 = T_0$ and $\beta_0 = 0$; (b) in the plane (E, β_0) of the initial energy and temporal chirp. Predictions of the variational approximation are marked as follows: the stability region is unshaded, while ones where the pulse is predicted to be unstable due to spreading out or collapse are shaded, respectively, by gray and dark gray. The numbered points are those at which direct simulations of Eq. (9.1) were performed, to verify the predictions, as explained in the text.

9.4   The three-dimensional case

The 3D generalization of the model (9.1), with the 2D transverse Laplacian, $u_{xx} + u_{yy}$, instead of u_{xx}, was also investigated by means of the VA and in direct simulations, starting with the ansatz (9.6) in which x is replaced by the transverse radial coordinate $r = \sqrt{x^2 + y^2}$ [122]. Both the variational and fully numerical approaches yield a negative result: the 3D soliton *can never be stabilized* by means of the longitudinal DM. The addition of NLM, i.e., periodic (in z) alternation of the sign of the coefficient in front of the cubic term, as in Eqs. (7.1), (7.2), does not stabilize 3D solitons either.

A possibility to find stable 3D solitons is offered by a model of a bulk layered medium (instead of the planar waveguide considered above). It includes the same longitudinal DM as in the above 2D model, which is combined with a transverse (in direction y) quasi-1D lattice, in the form of periodic modulation of the RI (refractive index):

$$i\frac{\partial u}{\partial z} + \left[\frac{1}{2}\left(\frac{\partial^2}{\partial x^2} + \frac{\partial^2}{\partial y^2} + D(z)\frac{\partial^2}{\partial \tau^2}\right) + \varepsilon\cos(2y) + |u|^2\right]u = 0. \qquad (9.16)$$

Here ε is the strength of the transverse modulation (the modulation period is normalized to be π), and the DM map is taken in the symmetric form (cf. Eq. (9.2),

$$D(z) = \begin{cases} \overline{D} + D_{\mathrm{m}} > 0, \, 0 < z < L, \\ \overline{D} - D_{\mathrm{m}} < 0, \, L < z < 2L, \end{cases} \qquad (9.17)$$

Parameters of the map are fixed by rescaling to be $L \equiv 1$ and $D_{\mathrm{m}} \equiv 1$, while the PAD $\overline{D}$ is small. This 3D model was very recently investigated, but only in the framework of VA in work [123] (direct simulations will be reported later).

The variational ansatz was taken in the form (cf. Eq. (9.6))

$$\begin{aligned} u &= A(z)\exp\left\{i\phi(z) - \frac{1}{2}\left[\frac{x^2}{W^2(z)} + \frac{y^2}{V^2(z)} + \frac{\tau^2}{T^2(z)}\right] + \right.\\ &\left. + \frac{i}{2}\left[b(z)\,x^2 + c(z)\,y^2 + \beta(z)\,\tau^2\right]\right\}, \end{aligned} \qquad (9.18)$$

where V and c are the additional width and chirp in the direction of RI modulation. The standard VA procedure leads to the conservation of the energy, $E \equiv A^2WVT$, and evolution equations (cf. Eqs. (9.12) and (9.13))

$$W'' = \frac{1}{W^3} - \frac{E}{2\sqrt{2}W^2VT}, \qquad (9.19)$$

$$V'' = \frac{1}{V^3} - 4\varepsilon V\exp\left(-V^2\right) - \frac{E}{2\sqrt{2}WV^2T}, \qquad (9.20)$$

$$\left(\frac{T'}{D}\right)' = \frac{D}{T^3} - \frac{E}{2\sqrt{2}WVT^2}, \qquad (9.21)$$

supplemented by the same matching (boundary) condition (9.14) as in the 2D model, and also by the relations (cf. Eq. (9.11))

$$b = \frac{W'}{W}, \; c = \frac{V'}{V}, \beta = \frac{T'}{D(z)T}. \tag{9.22}$$

For large ε (a strong lattice), one may keep only the first two terms on the right-hand side of Eq. (9.20). This approximation yields a nearly constant value V_0 of V, which is a smaller root of the corresponding equation,

$$4\varepsilon V_0^4 \exp\left(-V_0^2\right) = 1 \tag{9.23}$$

(a larger root corresponds to an unstable equilibrium of Eq. (9.20)). The roots exist provided that

$$\varepsilon > \varepsilon_{\min} = e^2/16 \approx 0.46, \tag{9.24}$$

the relevant one being limited by $V_0 < 2$. Then, the substitution of $V = V_0$ in the remaining equations (9.19) and (9.21) leads to essentially the same VA-generated dynamical system for the 2D model which was shown in the previous section to give rise to stable STSs.

Simulations of Eqs. (9.19)-(9.21) readily produce solutions corresponding to stable 3D solitons, an example of which is shown in Fig. 9.5. The analytical prediction (9.24) for the minimum RI modulation amplitude, necessary for the existence of stable solitons, is very accurately verified by numerical data. The simulations also reveal finite minimum and maximum values of the energy which border the area of stable solitons. It is noteworthy too that, as well as in the case of the ordinary DM solitons in optical fibers (see Fig. 2.5), stable 3D solitons in the present model can be predicted at small *negative* values of $\overline{D}$, up to $\left(-\overline{D}\right)_{\max} \approx 0.005$, which corresponds to normal (rather than anomalous) average GVD in the system.

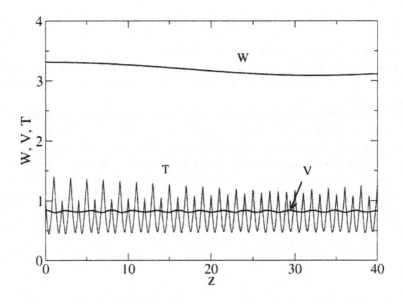

Figure 9.5: An example of the stable evolution of solutions to the variational equations (9.19)-(9.21), which approximate 3D solitons in the model (9.16) combining the longitudinal DM and periodic modulation of the refractive index in one transverse direction. The soliton's widths in the direction x, y and τ, i.e., W, V and T, are shown as functions of z, for $E = 0.5$, $\varepsilon = 1$, and $\overline{D} = 0$.

Chapter 10

Feshbach-resonance management in optical lattices

10.1   Introduction to the topic

Stabilization of 3D solitons in nonlinear optics and BECs (Bose-Einstein condensates) is a cardinal problem in the current studies in these fields. Various theoretical approaches to this topic constitute one of central themes of the present book. As explained in previous chapters, the ac-FRM technique, in the form of periodic reversal of the sign in front of the nonlinear terms in the corresponding GPE (Gross-Pitaevskii equation), is sufficient to stabilize only 2D solitons in BEC. On the other hand, a quasi-1D optical lattice (OL), which is much easier to create in the experiment than its multi-dimensional (2D or 3D) counterparts, is also capable to support stable solitons only in the 2D setting, but not in three dimensions [26]. These findings suggest a natural question, whether a combination of FRM and quasi-1D lattice may be sufficient to stabilize 3D solitons. In a recent work [166] a positive answer was given to this question, using an analytical approach (VA) and direct simulations.

The interplay of an OL and *low-frequency* FRM suggests another interesting possibility. Indeed, it is well known that the GPE equipped with the OL gives rise to regular solitons or GSs (gap solitons) if the nonlinearity is, respectively, self-attractive or self-repulsive. Then, in the case of periodic slow switching between the two signs of the nonlinearity provided by the FRM, a question arises, whether periodic adiabatic transitions between solitons of the two types can be predicted. The corresponding stable *alternate solitons* are possible indeed, as was demonstrated in both the 2D and 1D settings [77]. An account of these results is also included in the present chapter.

10.2 Stabilization of three-dimensional solitons by the Feshbach-resonance management in a quasi-one-dimensional lattice

10.2.1 The model and variational approximation

The model introduced in work [166] is based on the GPE in three dimensions, that includes the 1D lattice potential and the same ac-FRM modulation of the nonlinearity coefficient as in other models considered in this book. Thus, the normalized equation for the single-particle wave function ψ is

$$i\frac{\partial \psi}{\partial t} = \left[-\frac{1}{2}\nabla^2 + \varepsilon\left(1 - \cos(2z)\right) + \left(g_0 + g_1 \sin(\Omega t)\right)|\psi|^2\right]\psi, \qquad (10.1)$$

where the Laplacian (kinetic-energy operator) ∇^2 acts on all the three coordinates x, y, and z. Further, ε is the strength of the OL potential, whose period is scaled to be π, g_0 and g_1 account for the dc and ac parts of the FRM-controlled nonlinearity coefficient, and Ω is the ac-FRM frequency. Solitons can be constructed only in the case of $g_0 < 0$, which implies that the constant part of the interaction is self-attractive.

Equation (10.1) can be derived from the Lagrangian,

$$L = \pi \int_{-\infty}^{+\infty} dz \int_0^\infty \varrho\, d\varrho \left[i\left(\frac{\partial \psi}{\partial t}\psi^* - \frac{\partial \psi}{\partial t}\psi\right) - \left|\frac{\partial \psi}{\partial \varrho}\right|^2 - \left|\frac{\partial \psi}{\partial z}\right|^2\right.$$
$$\left. -2\varepsilon\left(1 - \cos(2z)\right)|\psi|^2 - \left(g_0 + g_1 \sin(\Omega t)\right)|\psi|^4\right], \qquad (10.2)$$

where the asterisk stands for the complex conjugation, and $\varrho \equiv \sqrt{x^2 + y^2}$ is the radial variable in the plane transverse to the lattice. To apply the VA to this model, a complex Gaussian ansatz may be adopted (cf., for instance, ansatz (9.6), with the real amplitude $A(t)$, phase $\phi(t)$, radial and axial widths $W(t)$ and $V(t)$, and the corresponding chirps $b(t)$ and $\beta(t)$:

$$\psi(\mathbf{r}, t) = A \exp\left[i\phi - \varrho^2\left(\frac{1}{2W^2} + ib\right) - z^2\left(\frac{1}{2V^2} + i\beta\right)\right]. \qquad (10.3)$$

An effective Lagrangian is obtained by inserting the ansatz (10.3) into Eq. (10.2). The Euler-Lagrange equations derived from the effective Lagrangian yield, first, the conservation of the solution's norm ($dE/dt = 0$), which is proportional to the number of atoms in the condensate,

$$E \equiv \frac{2}{\sqrt{\pi}} \int_{-\infty}^{+\infty} dz \int_0^\infty \varrho\, d\varrho |\psi|^2 = A^2 W^2 V, \qquad (10.4)$$

and dynamical equations for the widths W and V (cf. Eqs. (9.13), (9.12), derived in a similar situation in the previous chapter):

$$\frac{d^2 W}{dt^2} = \frac{1}{W^3} + \frac{E}{\sqrt{8}W^3 V}\left[g_0 + g_1 \sin(\Omega t)\right], \qquad (10.5)$$

$$\frac{d^2 V}{dt^2} = \frac{1}{V^3} - 4\varepsilon V \exp\left(-V^2\right) + \frac{E}{\sqrt{8}W^2 V^2}\left[g_0 + g_1 \sin(\Omega t)\right]. \qquad (10.6)$$

Numerical results will be given below for the normalization $E = \pi^{-3/2}$ (this condition may always be imposed by rescaling g_0 and g_1).

A necessary condition for the existence of a 3D soliton in the present model can be derived from these equations in an approximate form. To this end, it is assumed that g_1 is small, while Ω is large in Eq. (10.1). It is also conjectured that the average value $\overline{W}$ of the soliton's radial size, W, is large (see below). Further, in the lowest approximation, the soliton's size in the axial direction may be assumed constant, $V(t) \approx V_0$, as determined by the relation

$$4\varepsilon V_0^4 \exp\left(-V_0^2\right) = 1, \tag{10.7}$$

that follows from Eq. (10.6) where the last small term ($\sim W^{-2}$) is dropped. Equation (10.7) has real solutions if the OL strength exceeds a minimum (threshold) value,

$$\varepsilon \geq \varepsilon_{\text{thr}} = e^2/16 \approx 0.46 \tag{10.8}$$

(note the same value appeared in a similar context in the model considered in the previous chapter, see Eq. (9.24)). For $\varepsilon \geq \varepsilon_{\text{thr}}$, Eq. (10.7) has two real solutions, which implies the existence of two different solitons. It seems very plausible (cf. the situation for static models considered Refs. [24, 25, 26]) that the narrower soliton, corresponding to smaller V_0, is stable, and the other one is unstable.

Next, replacing $V(t)$ by V_0 in Eq. (10.6), one may look for a solution as $W(t) \approx \overline{W} + W_1 \sin(\Omega t)$. For large Ω, the variable part of the equation yields

$$W_1 = -\frac{Eg_1}{\sqrt{8\overline{W}^3} V_0 \Omega^2}. \tag{10.9}$$

Then, the consideration of the constant part of Eq. (10.5), with regard to the first correction generated by averaging of the product of $W_1 \sin(\Omega t) g(t)$, yields the following result:

$$\overline{W}^4 = \frac{3}{4\sqrt{2V_0}} \left(\frac{Eg_1}{\Omega}\right)^2 \frac{1}{\left(E|g_0| - \sqrt{8V_0}\right)}. \tag{10.10}$$

An essential corollary of Eq. (10.10) is a necessary condition for the existence of the 3D soliton:

$$|g| > (|g_0|)_{\text{min}} = \frac{\sqrt{8V_0}}{E} \tag{10.11}$$

(recall g_0 is negative). In fact, this minimum value, for given E, corresponds (within the framework of the VA) to the critical norm necessary for the existence of the 2D soliton (i.e., the norm of the TS). Direct simulations presented in the next subsection show that this condition holds indeed, albeit approximately.

Finally, the above conjecture, that $\overline{W}^4$ is large, is (formally) valid only when $|g_0|$ slightly exceeds the minimum value defined in Eq. (10.11) – then, the expression (10.10) will be large, as the denominator is small.

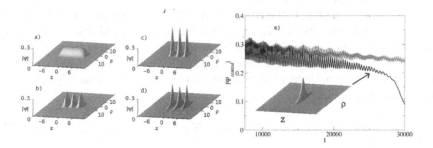

Figure 10.1: Evolution of $|\psi(x, y, z, t)|$ in the numerical experiment providing for the formation of a three-peak soliton in the 3D model combining the ac-FRM management and quasi-1D lattice (the final form of the model is given by Eq. (10.1)). Panels a) through d) display the shape of the wave functions at different moments of time. The established soliton in panel d) corresponds to $g_0 = -18$, $g_1 = 4g_0$, $\varepsilon = 20.5$, and $\Omega = 22$. In panel (e), thin and thick lines show the evolution of the amplitude of the central peak in two different cases: the three-peaked soliton proper, and in the case when two side peaks were suddenly removed (the latter configuration is shown in the inset).

10.2.2   Numerical results

PDE simulations of the full model (Eq. 10.1) produce stable single-peaked solitons (as predicted by the VA) and their multi-peaked counterparts. However, it is difficult to enforce direct self-trapping of an initial Gaussian pulse into the soliton, therefore, a special procedure was worked out in work [166] to construct stable solitons in this model. To this end, simulations began with Eq. (10.1) that contained additional terms (an *ad hoc* potential providing for initial trapping in the ϱ and z directions); the coefficients in front of other terms were also different from what is written in Eq. (10.1). Then, the added terms were gradually removed, and coefficients in front of the remaining terms were cast in the final form, corresponding to Eq. 10.1.

Figure 10.1 displays self-trapping of a typical three-peaked soliton in the numerical experiment. Panel (d) in the figure shows the established pattern, which then remains unchanged over indefinitely long time. In all stable multi-peaked solitons, the relative phase difference between adjacent peaks is close to π.

It is important to understand whether the multi-peak solitons are true coherent bound states, supported by the interaction between peaks, or just sets of quasi-2D solitons, completely isolated from each other by barriers in the strong OL potential. To this end, panel (e) in Fig. 10.1 shows the evolution of the central-peak's amplitude in the three-peaked soliton itself, and in the case when two side peaks were suddenly eliminated. As is seen, in the former case the amplitude performs oscillations without systematic decay, while decay begins if the central peak is no longer supported by the

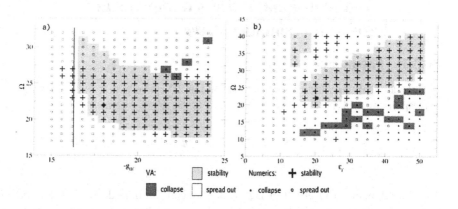

Figure 10.2: Stability regions for the 3D solitons in the model combining the FRM and quasi-1D lattice, as predicted by the variational approximation ("VA"), and as found from direct simulations of the Gross-Pitaevskii equation (10.1) ("numerics"). a) The (g_0, Ω) parameter plane; b) the (ϵ, Ω) plane (in this figure, g_0 and ε are denoted, respectively, as g_{0f} and ε_f). Other parameters are as in Fig. 10.1. The vertical line in (a) corresponds to the minimum value of $|g_0|$ predicted by Eq. (10.11), which actually corresponds to the two-dimensional Townes soliton. The fat dot in panel (b) (at $\varepsilon \approx 20.5, \Omega \approx 22$) corresponds to the example displayed in Fig. 10.1.

interaction with the side ones. Eventually the single peak completely decays in this case (although stable single-peak solitons can be found in this model too).

If the strength of the OL is increased by a factor $\gtrsim 2$ against the value for which Fig. 10.1 was generated, the stable multi-peaked pattern becomes a set of virtually uncoupled fundamental solitons, each being trapped in a single lattice cell (then, in contrast with what was shown in Fig. 10.1(e), the removal of matter from any subset of the cells does not affect the localized states in other cells in any tangible way). In fact, in the latter case one is dealing with a nearly 2D situation, as the condensate is tightly confined by individual potential wells in the OL. It was known before that a very strong 1D parabolic potential can stabilize ac-FRM-driven solitons in three dimensions, just by making them effectively two-dimensional objects [130]

Results of systematic direct simulations of the GPE (10.1) are collected and compared with predictions of the VA (which follow from simulations of ODEs (10.5) and 10.6) in Fig. 10.2. As is seen, the agreement between the VA and full numerical results is quite good.

The variational estimate (10.11) for the minimum size of the average nonlinear coefficient necessary for the existence of the 3D soliton in the ac-FRM-driven Q1D lattice is borne out by direct simulations, although approximately. Figure 10.2(b) confirms the existence of the minimum OL strength ε which is necessary to support the 3D solitons, as predicted by the VA, see Eq. (10.8).

10.3   Alternate regular-gap solitons in one- and two-dimensional lattices under the ac Feshbach-resonance drive

10.3.1   The model

The model dealt with in this section is set in the 2D space with a full 2D lattice. The corresponding GPE is

$$i\frac{\partial\psi}{\partial t} + \frac{\partial^2\psi}{\partial x^2} + \frac{\partial^2\psi}{\partial y^2} + \varepsilon\left[\cos(2x) + \cos(2y)\right]\psi + \left[\lambda_0 + \lambda_1\cos(\omega t)\right]|\psi|^2\psi = 0,$$

(10.12)

The only dynamical invariant of Eq. (10.12) (with the time-dependent nonlinearity coefficient) is the norm, which is proportional to the number of atoms in the condensate,

$$N = \int\int |\psi(x, y, t)|^2\, dxdy.$$

(10.13)

The objective of the consideration is to find *alternate solitons*, which perform periodic adiabatic transitions between gap-soliton configurations and regular solitons that correspond, respectively, to $\lambda_0 + \lambda_1\cos(\omega t) < 0$ and $\lambda_0 + \lambda_1\cos(\omega t) > 0$, in the case when the modulation frequency ω is small enough. The results included in this section were obtained in work [77].

In case of $\lambda_1 = 0$, stationary solutions to Eq. (10.12) are sought as

$$\psi(x, y, t) = u(x, y)\exp(-i\mu t),$$

(10.14)

with a real chemical potential μ and a real function u which satisfies the equation

$$\mu u + \frac{\partial^2 u}{\partial x^2} + \frac{\partial^2 u}{\partial y^2} + \varepsilon\left[\cos(2x) + \cos(2y)\right]u + \lambda_0 u^3 = 0.$$

(10.15)

Search for soliton solutions should be preceded by consideration of the spectrum of the linearized version of Eq. (10.15), as solitons may only exist at values of μ belonging to *bandgaps* in the spectrum. The linearization of Eq. (10.15) leads to a separable 2D eigenvalue problem,

$$\left(\hat{L}_x + \hat{L}_y\right)u(x, y) = -\mu u(x, y),$$

(10.16)

where a 1D linear operator is $\hat{L}_x \equiv \partial^2/\partial x^2 + \varepsilon\cos(2x)$. The corresponding eigenstates can be built as $u_{kl}(x, y) = g_k(x)g_l(y)$, with the eigenvalues $\mu_{kl} = \mu_k + \mu_l$, where $g_k(x)$ and $g_l(y)$ is any pair of quasi-periodic Bloch functions solving the linear ME (Mathieu equation), $\hat{L}_x g_k(x) = -\mu_k g_k(x)$, μ_k and μ_l being the corresponding eigenvalues. The band structure of the 2D linear equation (10.16) constructed this way was already investigated in detail (see, e.g., papers [135] and [56]). It includes, as the spectrum of the ME itself, a semi-infinite gap which extends to $\mu \rightarrow -\infty$, and a

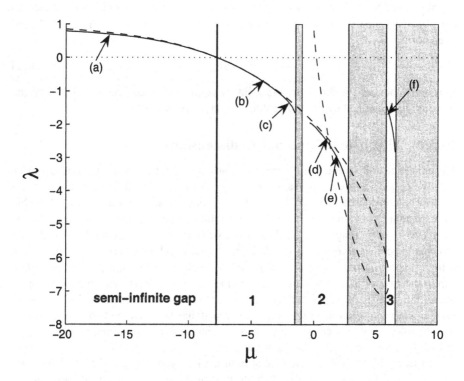

Figure 10.3: The system of vertical stripes gives a typical example (for $\varepsilon = 7.5$) of the bandgap structure found from the linearized version of Eq. (10.15), shaded and unshaded zones being the Bloch bands and gaps, respectively (solitons may exist only inside the gaps). The solid curve shows the dependence $\mu(\lambda_0)$ for numerical soliton solutions, as found (also for $\varepsilon = 7.5$, and for a fixed norm, $N = 4\pi$) from the full nonlinear stationary equation (10.15). The dashed curve is the same dependence for approximate soliton solutions found by means of a variational method based on a Gaussian ansatz (details of the latter are given in paper [77]). Points "a" through "f" mark particular solitons whose shapes are given in Fig. 10.4.

set of finite gaps separated by bands that are populated by quasiperiodic Bloch-wave solutions, see a typical example in Fig. 10.3.

With the constant self-attractive nonlinearity ($\lambda_1 = 0, \lambda_0 > 0$), a family of stable stationary 2D solitons is known to exist in the semi-infinite gap [24, 174, 56]. With the repulsive constant nonlinearity, $\lambda_0 < 0$, stable 2D gap-soliton solutions can be found in finite gaps [23, 135, 26, 152]. In either case, a necessary condition for the existence of the stationary 2D solitons is that their norm, defined by Eq. (10.13), must exceed a certain minimum (*threshold*) value, N_{thr} [56].

Stationary soliton families are characterized by the corresponding dependences $\mu(\lambda_0)$ in each gap where the solitons exist. For a fixed OL strength $\varepsilon = 7.5$, and fixed norm,

$$N = 4\pi, \tag{10.17}$$

these dependences are shown by the solid curve in Fig. 10.3. Possible shapes of the solitons are illustrated by a set of profiles displayed in Fig. 10.4.

10.3.2   Alternate solitons in two dimensions

The existence and stability of alternate solitons were studied by direct simulations of Eq. (10.12). The case of the vanishing dc part in the nonlinearity coefficient, $\lambda_0 = 0$, is to be considered first. At $t = 0$, the simulations started with an initial profile corresponding to a numerically found soliton solution of the stationary equation (10.15) with $\lambda = \lambda_1$ (assuming $\lambda_1 > 0$). Systematic simulations demonstrate that it is indeed possible to achieve *stable* periodic adiabatic alternations between two quasi-stationary soliton shapes, one corresponding to an ordinary soliton belonging to the semi-infinite gap in the case of the constant attractive nonlinearity, the other being a gap soliton in one of the finite gaps, supported by the constant repulsive nonlinearity. Relaxation to such an *alternate soliton*, which periodically switches between the two limit forms, is accompanied by very weak radiation loss. In the established regime, no emission of radiation could be detected in the simulations.

An example of a robust alternate soliton is displayed in Fig. 10.5. In particular, sidelobes in the soliton's profile, characteristic of the gap-soliton shape, periodically appear and disappear in the course of the cyclic evolution. It is noteworthy that periodic crossings of the zero-nonlinearity point, $\lambda = 0$, at which no stationary soliton may exist, do not destroy the alternate soliton. The spatially-averaged squared width of the soliton, the evolution of which is shown in the lower panel of the figure, is defined as

$$\xi^2(t) \equiv \frac{\int\int x^2 |u(x,y,t)|^2 dx dy}{\int\int |u(x,y)|^2 dx dy}. \tag{10.18}$$

Results of systematic simulations are summarized in stability diagrams for the alternate solitons, which are displayed in Fig. 10.6 for $\lambda_0 = 0$ and several different values of the FRM amplitude λ_1. Naturally, the solitons may be stable under low-frequency (quasi-adiabatic) FRM drive. The stability region in Fig. 10.6 is defined as one in which the total radiation loss, measured in the course of indefinitely long evolution, is less than 2% of the initial norm. In particular, the example shown in Fig. 10.5

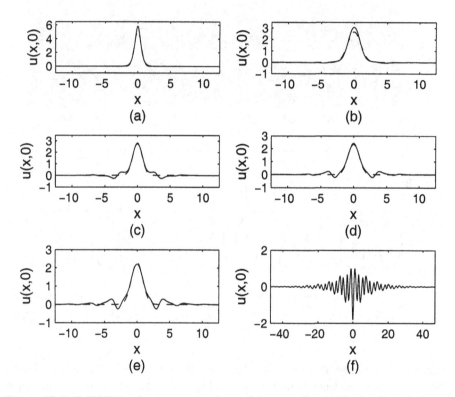

Figure 10.4: Solid lines show cross sections of the two-dimensional solitons' shapes through the line $y = 0$, i.e., $u(x, 0)$, see Eq. (10.14. Panels (a) to (f) correspond to the string of marked points in Fig. 10.3.

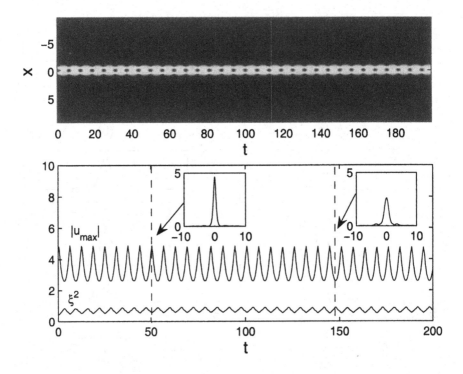

Figure 10.5: An example of a stable alternate soliton, for $\varepsilon = 5$, $\lambda_1 = 0.7$, $\omega = 1$, and $\lambda_0 = 0$; it corresponds to point (a) in Fig. 10.6 (for $\lambda_1 = 0.7$). The upper panel displays the soliton evolution in terms of contour plots. The lower panel shows the time dependence of the amplitude and mean squared width of the soliton. Two insets are cross-sections of instantaneous profiles of the soliton taken at moments of time ($t = 50$ and $t = 150$) when it is very close to a regular soliton and gap soliton, respectively.

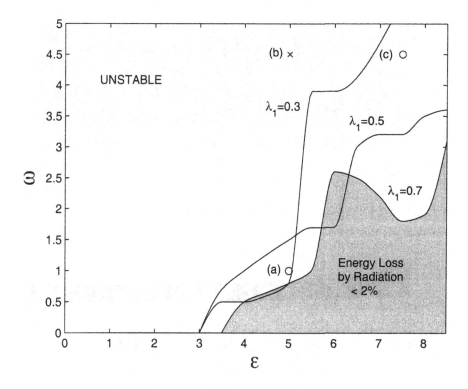

Figure 10.6: Stability diagram for the alternate solitons in the (ε,ω) plane for $\lambda_0 = 0$ and different fixed values of the Feshbach-resonance-management amplitude λ_1. The region of complete stability (it is shaded for $\lambda_1 = 0.7$) is defined so that the total radiation loss of the soliton's initial norm is less than 2% in this region.

corresponds to the point (a) in Fig. 10.6 (for $\lambda_1 = 0.7$); in this case, the total loss is almost exactly 2%.

As the driving frequency ω increases, the soliton emits more radiation. For moderately high frequencies, the initial solitary-wave pulse prepared as said above (i.e., as a numerically exact stationary soliton corresponding to the initial value of λ) sheds off a conspicuous share of its norm; then, the emission of radiation ceases, and the remaining part of the pulse self-traps into a robust alternate soliton. An example of a such a *semi-stable* dynamical regime is displayed in Fig. 10.7. To additionally illustrate the relaxation to the stable regime, the lower panel of the figure includes a curve $\rho(t)$ which shows the evolution of the soliton's norm in time. In this case, the resulting alternate soliton oscillates between nearly stationary shapes corresponding to points (a) and (b) in Fig. 10.3, which belong to the semi-infinite and first finite gaps, respectively (there-

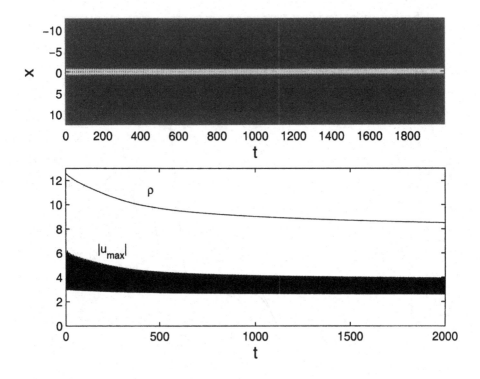

Figure 10.7: An example of a "semi-stable" soliton, for $\varepsilon = 7.5, \lambda_1 = 0.7, \omega = 4.5$, and $\lambda_0 = 0$, which corresponds to point (c) in Fig. 10.6 (for $\lambda_1 = 0.7$). After 1400 oscillation periods, the soliton definitely survives.

fore, the former one corresponds to regular solitons, and the latter one – to solitons of the gap type). In general, the alternate solitons become more robust with the increase of the OL strength ε.

In Fig. 10.6, the semi-stable regimes are not marked separately from completely unstable ones, as the border between them is fuzzy (in particular, it is not quite clear whether the semi-stable solitons would not very slowly decay on an extremely long time scale). In any case, a broad area adjacent to the completely stable one in Fig. 10.6 is actually a region of semi-stability. At still higher driving frequencies, the soliton is definitely destroyed.

It is not quite clear either if the stability region of the alternate solitons is limited on the side of very small frequencies. Indeed, one may expect that, in this case, the soliton spending long time around the zero-nonlinearity point, $\lambda = 0$, will spread out, and may thus decay; on the other hand, if the soliton is very broad by itself, it may survive this temporary spreading out. The lowest frequency checked in the simulations was $\omega = 0.1$, the solitons being unequivocally stable at this point.

It is also relevant to address the issue of the existence of the threshold necessary for the formation of the soliton, which, as mentioned above, exists for both $\lambda > 0$ and $\lambda < 0$ in the static 2D models. In the nonstationary (FRM-driven) model with the fixed norm (see Eq. (10.17)), rescaling (necessary to keep the norm fixed) shows that the threshold manifests itself in the fact that persistent alternate solitons cannot be found if the FRM-drive's amplitude is too small, $\lambda_1 < (\lambda_1)_{\text{thr}}$. In the strong lattice with $\varepsilon = 7.5$ (the situation in the stationary model at this value of ε is illustrated by Figs. 10.3 and 10.4), the threshold exists but is so small that its accurate value cannot be identified. This is possible for smaller ε. In particular, for $\varepsilon = 4$ it was found to be $(\lambda_1)_{\text{thr}} \approx 0.15$, which should be compared to the the thresholds for the same $\varepsilon = 4$ and the same fixed norm (10.17) for the ordinary and gap solitons in the corresponding static 2D models: $\lambda_{\text{thr}}^{(\text{ord})} \approx 0.04$, and $\lambda_{\text{thr}}^{(\text{gap})} \approx -0.04$, respectively. Quite naturally, the dynamical threshold is much higher than its static counterparts.

The FRM-driven soliton dynamics with a negative nonzero dc part of the nonlinearity coefficient, $\lambda_0 < 0$, which corresponds to repulsion, was also considered. The interaction being repulsive on average, one may expect the existence of gap solitons in the high-frequency limit. The situation may be illustrated by an example for $\lambda_0 = -0.9$ and $\lambda_1 = 1.6$. The simulations started with the initial profile corresponding to point (a) in Fig. 10.3, as the initial value of the nonlinearity coefficient, $\lambda(0) = 0.7$, pertains to this point. The minimum instantaneous value of the oscillating nonlinear coefficient is $\lambda_{\min} = -2.5$ in this case, the stationary solution with $\lambda = -2.5$ pertaining to point (d) in Fig. 10.3, which belongs to the second finite band. In this regime, the oscillating nonlinear coefficient $\lambda(t)$, in addition to cycling across the point $\lambda = 0$ and a very narrow Bloch band separating the semi-infinite and first finite gaps, periodically passes the wider Bloch band between the first and the second finite gaps, where stationary solitons cannot exist. Nevertheless, a stable alternate soliton, found in this case, survives all the traverses of the "dangerous zones", as shown in Fig. 10.8. A small amount of radiation is emitted at the initial stage of the evolution (the curve $\rho(t)$ again shows the evolution of the soliton's norm), and then a robust alternate soliton establishes itself, cf. Fig. 10.7. It is noteworthy that, as seen in the insets, this soliton develops sidelobes that actually do not oscillate together with its core, and do not disappear either as λ takes positive values. The latter feature distinguishes this stable regime from the one shown in Fig. 10.5, where the sidelobes periodically disappear.

Finally, the application of the FRM mechanism to weakly localized ("loosely-bound") solitons, such as the one in panel (f) of Fig. 10.4, cannot produce any stable regime periodically passing through a shape of this type, for any combination of λ_0 and λ_1 in Eq. (10.12).

10.3.3 Dynamics of one-dimensional solitons under the Feshbach-resonance management

The 1D case also deserves consideration, as the experiment may be easier in this case, and it is interesting to compare the results with those reported above for the 2D model.

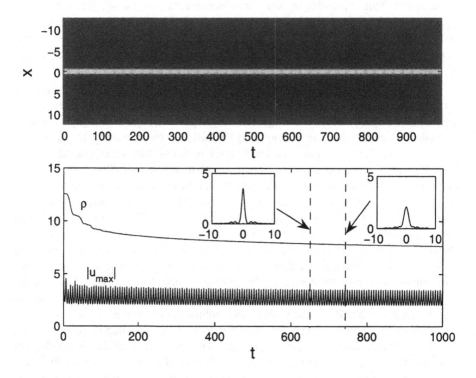

Figure 10.8: A stable alternate soliton for $\epsilon = 7.5$, $\lambda_0 = -0.9$, $\lambda_1 = 1.6$, and $\omega = 1$. The soliton survives while the nonlinear coefficient periodically traverses both the $\lambda = 0$ point and two Bloch bands separating the semi-infinite and first two finite bands. Insets show cross-sections of instantaneous soliton's profiles with the largest (left) and smallest (right) amplitudes.

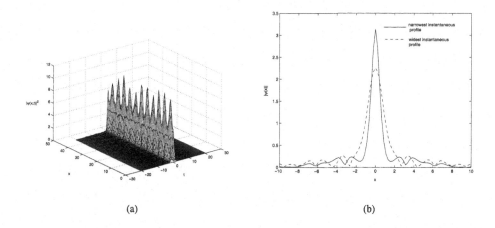

(a)   (b)

Figure 10.9: (a) A typical example of a stable alternate soliton found in the one-dimensional model (10.19), for $\varepsilon = 4.5$, $\lambda_1 = 1$, $\omega = \pi/2$, and $\lambda_0 = 0$. (b) Two profiles between which the alternate soliton periodically oscillates.

In particular, an issue is whether the existence of the 1D alternate soliton requires any threshold condition (minimum norm). The effective one-dimensional GPE is a straightforward reduction of Eq. (10.12),

$$i\frac{\partial \psi}{\partial t} + \frac{\partial^2 \psi}{\partial x^2} + \varepsilon \cos(2x)\psi + [\lambda_0 + \lambda_1 \cos(\omega t)]\,|\psi|^2\psi = 0. \qquad (10.19)$$

Here, results will be presented only for the most fundamental case without the dc component in $\lambda(t)$, i.e., $\lambda_0 = 0$, and fixing $|\lambda_1| = 1$. The strategy is the same as in the 2D case: direct simulations of Eq. (10.19) start with a soliton profile that would be a numerically exact stationary soliton for the initial value of the nonlinearity coefficient, $\lambda = \lambda(0)$. In most cases, the solution's 1D norm was fixed at $N \equiv \int_{-\infty}^{+\infty} |u(x)|^2\,dx = 7.9$ (this normalization was chosen as it corresponds to an almost constant value of the chemical potential, $\mu \approx 2$, in the stationary version of the problem). However, the overall stability diagram will include different values of N, see Fig. 10.10 below.

In the 1D case, stable alternate solitons can be readily found, see an example in Fig. 10.9(a). The soliton periodically oscillates between the narrow and wide profiles, which are displayed in Fig. 10.9(b).

The stability diagram for the 1D alternate solitons, based on systematic numerical simulations, is displayed in Fig. 10.10. It is noteworthy that the shape of the stability area is qualitatively similar to that in the 2D version of the model, cf. Fig. 10.6. The similarity suggest that the basic results for the stability of the alternate solitons in the FRM-driven model with the OL potential are quite generic.

It is well known that the existence of the ordinary and gap solitons *does not* require any finite threshold (minimum norm) in the static 1D models with both $\lambda > 0$ and

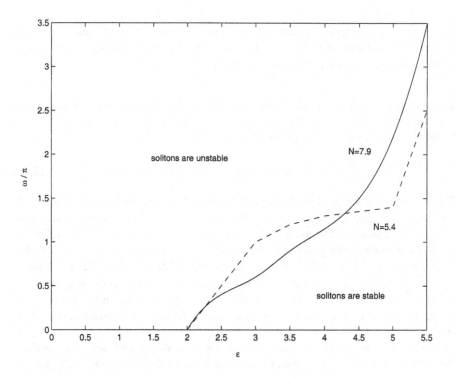

Figure 10.10: Stability diagram for alternate solitons in the one-dimensional model (10.19) combining the ac-FRM drive, with $\lambda_0 = 0$, and lattice potential. The stability borders are shown for two different values of the fixed norm.

$\lambda < 0$. A principal difference of the dynamic (FRM-driven) 1D model is that persistent alternate solitons can be found only *above a finite threshold*, $N > N_{\text{thr}}$. For instance, the numerical results yield $N_{\text{thr}} \approx 2.1$ for $\varepsilon = 7.5$ and $\omega = \pi/2$. Thus, in this sense the 1D dynamic model is closer to the 2D one than to its static 1D counterparts.

Besides the fundamental (single-peaked) 1D solitons considered above, stable higher-order (multi-peaked) alternate solitons can also be found in the FRM-driven 1D model. As concerns static higher-order solitons on lattices, a known example is the so-called twisted localized mode, i.e., an odd (antisymmetric) soliton in the discrete NLS equation [45]. A similar object is a bound state of two lattice solitons, which, too, is stable only in the anti-symmetric configuration, in the 1D [84] and 2D [89] cases alike.

Following the pattern of the static lattice models, an antisymmetric initial state was prepared as a superposition of two separated stationary solitons with opposite signs (i.e., the phase difference π), corresponding to the initial value of λ. Direct simulations demonstrate that stable alternate antisymmetric solitons can be easily found this way, see a typical example in Fig. 10.11. Stable bound states of several fundamental solitons with the phase shift π between them were found too in the present model.

A 2D counterpart of the odd soliton would be a vortical soliton. Such stable vortices were found indeed in the static models with self-attraction [24, 174] and self-repulsion [26, 152, 136]. However, simulations have not produced stable solitons with intrinsic vorticity in the 2D FRM-driven model based on Eq. (10.12) with $\lambda_0 = 0$.

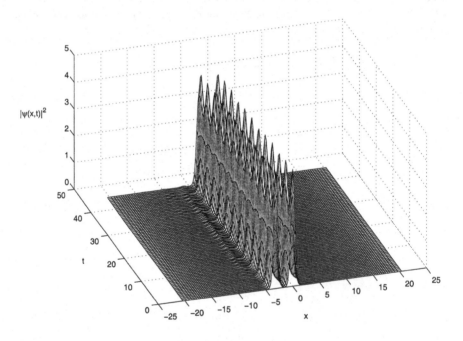

Figure 10.11: An example of a stable odd (antisymmetric) one-dimensional alternate soliton, found for $\lambda_1 = -1$, $\varepsilon = 5$, $\omega = \pi/2$, and $\lambda_0 = 0$.

Bibliography

[1] F. Kh. Abdullaev and B. B. Baizakov, *Disintegration of a soliton in a dispersion-managed optical communication line with random parameters*, Opt. Lett. **25**, 93 (2000).

[2] F. Kh. Abdullaev, B. B. Baizakov, and M. Salerno, *Stable two-dimensional dispersion-managed soliton*, Phys. Rev. E **68**, 066605 (2003).

[3] F. Kh. Abdullaev and J. G. Caputo, *Validation of the variational approach for chirped pulses in fibers with periodic dispersion*, Phys. Rev. E **58**, 6637 (1998).

[4] F. Kh. Abdullaev, J. G. Caputo, and N. Flytzanis, *Envelope soliton propagation in media with temporally modulated dispersion*, Phys. Rev. E **50**, 1552 (1994).

[5] F. Kh. Abdullaev, J. G. Caputo, R. A. Kraenkel, and B. A. Malomed, *Controlling collapse in Bose-Einstein condensation by temporal modulation of the scattering length*, Phys. Rev. A **67**, 013605 (2003).

[6] F. Kh. Abdullaev and R. Galimzyanov, *The dynamics of bright matter wave solitons in a quasi-one-dimensional Bose–Einstein condensate with a rapidly varying trap*, J. Phys. B **36**, 1099 (2003).

[7] F. Kh. Abdullaev and J. Garnier, *Collective oscillations of one-dimensional Bose-Einstein gas in a time-varying trap potential and atomic scattering length*, Phys. Rev. A **70**, 053604 (2004).

[8] F. Kh. Abdullaev and P. K. Khabibullaev. *Dynamics of Solitons in Inhomogeneous Condensed Media* (Fan Publishers: Tashkent, 1986) (in Russian).

[9] M. J. Ablowitz and G. Biondini, *Multiscale pulse dynamics in communication systems with strong dispersion management*, Opt. Lett. Opt. Lett. **23**, 1668 (1998).

[10] M. J. Ablowitz, G. Biondini, S. Chakravarty, and R. L. Horne, *On timing jitter in wavelength-division multiplexed soliton systems*, Opt. Commun. **150**, 305 (1998).

[11] M. Ablowitz and H. Segur. *Solitons and the Inverse Scattering Transform* (SIAM: Philadelphia, 1981).

[12] M. Abramowitz and I. A. Stegun. *Handbook of Mathematical Functions* (Dover Publication: New York, 1965).

[13] A. B. Aceves and S. Wabnitz, *Self-induced transparency solitons in nonlinear refractive periodic media*, Phys. Lett. **141**, 37 (1989).

[14] S. K. Adhikari, *Stabilization of bright solitons and vortex solitons in a trapless three-dimensional Bose-Einstein condensate by temporal modulation of the scattering length*, Phys. Rev. A **69**, 063613 (2004).

[15] G. P. Agrawal. *Nonlinear Fiber Optics* (Academic Press: San Diego, 1995).

[16] G. P. Agrawal. *Nonlinear Optical Communication Networks* (John Wiley & Sons, Inc.: New York, 1997).

[17] O. Aharon, B. A. Malomed, Y. B. Band, and U. Mahlab, *Minimization of the pulse's timing jitter in a dispersion-compensated WDM system*, Opt. Quant. Electr. **36**, 349 (2004).

[18] J. S. Aitchison, A. M. Weiner, Y. Silberberg, M. K. Oliver, J. L. Jackel, D. E. Leaird, E. M. Vogel, and P. W. E. Smith, *Observation of spatial optical solitons in a nonlinear glass wave-guide*, Opt. Lett. **15**, 471 (1990).

[19] D. Anderson, *Variational approach to nonlinear pulse propagation in optical fibers*, Phys. Rev. A **27**, 3135-3145 (1983).

[20] J. Atai and B. A. Malomed, *Families of Bragg-grating solitons in a cubic-quintic medium*, Phys. Lett. A **155**, 247 (2001).

[21] J. Atai and B. A. Malomed, *Spatial solitons in a medium composed of self-focusing and self-defocusing layers*, Phys. Lett. A **298**, 140 (2002).

[22] B. Baizakov, G. Filatrella, B. Malomed, and M. Salerno, *A double parametric resonance for matter-wave solitons in a time-modulated trap*, Phys. Rev. E **71**, 036619 (2005).

[23] B. B. Baizakov, V. V. Konotop, and M. Salerno, *Regular spatial structures in arrays of Bose-Einstein condensates induced by modulational instability*, J. Phys. B: At. Mol. Opt. Phys. **35**, 5105 (2002).

[24] B. B. Baizakov, B. A. Malomed, and M. Salerno, *Multidimensional solitons in periodic potentials*, Europhys. Lett. **63**, 642 (2003).

[25] B. B. Baizakov, M. Salerno, and B. A. Malomed, *Multidimensional solitons and vortices in periodic potentials*, in: *Nonlinear Waves: Classical and Quantum Aspects*, ed. by F. Kh. Abdullaev and V. V. Konotop, p. 61 (Kluwer Academic Publishers: Dordrecht, 2004); also available at http://rsphy2.anu.edu.au/~asd124/Baizakov_2004_61_Nonlinear Waves.pdf.

[26] B. B. Baizakov, B. A. Malomed and M. Salerno, *Multidimensional solitons in a low-dimensional periodic potential*, Phys. Rev. A **70**, 053613 (2004).

[27] O. Bang, C. B. Clausen, P. L. Christiansen, and L. Torner, *Engineering competing nonlinearities*, Opt. Lett. **24**, 1413 (1999).

[28] I. V. Barashenkov, D. E. Pelinovsky, and E. V. Zemlyanaya, *Vibrations and oscillatory instabilities of gap solitons*, Phys. Rev. Lett. **80**, 5117 (1998).

[29] L. Bergé, *Wave collapse in physics: principles and applications to light and plasma waves*, Phys. Rep. **303**, 259-370 (1998).

[30] L. Bergé, V. K. Mezentsev, J. Juul Rasmussen, P. L. Christiansen, and Yu. B. Gaididei, *Self-guiding light in layered nonlinear media*, Opt. Lett. **25**, 1037 (2000).

[31] A. Berntson, N. J. Doran, W. Forysiak, and J. H. B. Nijhof, *Power dependence of dispersion-managed solitons for anomalous, zero, and normal path-average dispersion*, Opt. Lett. **23**, 900 (1998).

[32] A. Berntson and B. A. Malomed, *Dispersion-management with filtering*, Opt. Lett. **24**, 507 (1999).

[33] L. Brzozowski and E. H. Sargent, *Optical signal processing using nonlinear distributed feedback structures*, IEEE J. Quant. Electr. **36**, 550 (2000).

[34] S. Burger, K. Bongs, S. Dettmer, W. Ertmer, K. Sengstock, A. Sanpera, G. V. Shlyapnikov, and M. Lewenstein, *Dark solitons in Bose-Einstein condensates*, Phys. Rev. Lett. **83**, 5198-5201 (1999).

[35] A. V. Buryak, P. Di Trapani, D. V. Skryabin, and S. Trillo, *Optical solitons due to quadratic nonlinearities: from basic physics to futuristic applications*, Phys. Rep. **370**, 63 (2002).

[36] D. K. Campbell and M. Peyrard, *Solitary wave collisions revisited*, Physica D **18**, 47-53 (1986).

[37] D. K. Campbell, M. Peyrard, and P. Sodano, *Kink-antikink interactions in the double sine-Gordon equation*, Physica D **19**, 165-205 (1986).

[38] G. M. Carter, J. M. Jacob, C. R. Menyuk, E. A. Golovchenko, and A. N. Pilipetskii, *Timing-jitter reduction for a dispersion-managed soliton system: Experimental evidence*, Opt. Lett. **22**, 513 (1997).

[39] A. R. Champneys, B. A. Malomed, J. Yang, and D. J. Kaup, *"Embedded solitons": solitary waves in resonance with the linear spectrum*, Physica D **152-153**, 340 (2001).

[40] P. Y. P. Chen, P. L. Chu and B.A. Malomed, *An iterative numerical method for dispersion managed solitons*, Opt. Commun. **245**, 425 (2005).

[41] R. Y. Chiao, E. Garmire, and C. H. Townes, *Self-trapping of optical beams*, Phys. Rev. Lett. **13**, 479-482 (1964).

[42] D. N. Christodoulides and R. I. Joseph, *Slow Bragg solitons in nonlinear periodic structures*, Phys. Rev. Lett. **62**, 1746 (1989).

[43] C. B. Clausen, O. Bang, and Y. S. Kivshar, *Spatial solitons and induced Kerr effects in quasi-phase-matched quadratic media*, Phys. Rev. Lett. 78, 4749-4752 (1997).

[44] L.-C. Crasovan, B. A. Malomed, and D. Mihalache, *Spinning solitons in cubic-quintic media*, Pramana – Indian J. Phys. **57**, 1041- 1059 (2001).

[45] S. Darmanyan, A. Kobyakov, and F. Lederer, *Stability of strongly localized excitations in discrete media with cubic nonlinearity*, Zh. Eksp. Teor. Fiz. **113**, 1253 (1998) [English translation: JETP **86**, 682 (1998)].

[46] J. Denschlag, J. E. Simsarian, D. L. Fede, C. W. Clark, L. A. Collins, J. Cubizolles, L. Deng, E. W. Hagley, K. Helmerson, W. P. Reinhardt, S. L. Rolston, B. I. Schneider, and W. D. Phillips, *Generating solitons by phase engineering of a Bose-Einstein condensate*, Science **287**, 97-101 (2000).

[47] A. De Rossi, C. Conti, and S. Trillo, *Stability, multistability, and wobbling of optical gap solitons*, Phys. Rev. Lett. **81**, 85 (1998).

[48] M. Desaix, D. Anderson, and M. Lisak, *Variational approach to collapse of optical pulses*, J. Opt. Soc. Am. B **8**, 2082 (1991).

[49] P. Di Trapani, D. Caironi, G. Valiulis, A. Dubietis, R. Danielius, and A. Piskarskas, *Observation of temporal solitons in second-harmonic generation with tilted pulses*, Phys. Rev. Lett. **81**, 570 (1998).

[50] R. Driben and B. A. Malomed, *Split-step solitons in long fiber links*, Opt. Commun. **185**, 439 (2000).

[51] R. Driben and B. A. Malomed, *Suppression of crosstalk between solitons in a multi-channel split-step system*, Opt. Commun. **197**, 481 (2001).

[52] R. Driben, B. A. Malomed, and P. L. Chu. *Solitons in regular and random split-step systems*, J. Opt. Soc. Am. B **219**, 143 (2003).

[53] R. Driben, B. A. Malomed, and P. L. Chu, *Transmission of pulses in a dispersion-managed fiber link with extra nonlinear segments*, Opt. Commun. **245**, 227 (2005).

[54] R. Driben, B. A. Malomed, M. Gutin, and U. Mahlab, *Implementation of nonlinearity management for Gaussian pulses in a fiber-optic link by means of second-harmonic-generating modules*, Opt. Commun. **218**, 93 (2003).

[55] N. K. Efremidis, S. Sears, D. N. Christodoulides, J. W. Fleischer, and M. Segev, *Discrete solitons in photorefractive optically induced photonic lattices*, Phys. Rev. E **66**, 046602 (2002).

[56] N. K. Efremidis, J. Hudock, D. N. Christodoulides, J. W. Fleischer, O. Cohen, and M. Segev, *Two-Dimensional Optical Lattice Solitons*, Phys. Rev. Lett. **91**, 213906 (2003).

[57] B. J. Eggleton, R. E. Slusher, C. M. de Sterke, P. A. Krug, and J. E. Sipe, *Bragg Grating Solitons*, Phys. Rev. Lett. **76**, 1627 (1996).

[58] B. Eiermann, Th. Anker, M. Albiez, M. Taglieber, P. Treutlein, K.-P. Marzlin, and M. K. Oberthaler, *Bright Bose-Einstein Gap Solitons of Atoms with Repulsive Interaction*, Phys. Rev. Lett. **92**, 230401 (2004).

[59] H. S. Eisenberg, Y. Silberberg, R. Morandotti, A. R. Boyd, and J. S. Aitchison, *Discrete spatial optical solitons in waveguide arrays*, Phys. Rev. Lett. **81**, 3383 (1998).

[60] P. Emplit and J. P. Hamaide, F. Reynaud, C. Froehly and A. Barthelemy, *Picosecond steps and dark pulses through nonlinear single mode fibers*, Opt. Commun. **62**, 374 (1987).

[61] P. O. Fedichev, Yu. Kagan, G. V. Shlyapnikov, and J. T. M. Walraven, *Influence of nearly resonant light on the scattering length in low-temperature atomic gases*, Phys. Rev. Lett. **77**, 2913 (1996).

[62] C. Etrich, F. Lederer, B. A. Malomed, T. Peschel, and U. Peschel, *Optical solitons in media with a quadratic nonlinearity*, Progress in Optics **41**, 483-568 (E. Wolf, editor: North Holland, Amsterdam, 2000).

[63] B.-F. Feng and B. A. Malomed, *Bound states of solitons between close wavelength-separated channels in a dispersion-managed fiber-optic link*, Opt. Commun. **219**, 143 (2003).

[64] A. Ferrando, M. Zacarés, P. Fernández de Córdoba, D. Binosi, and J. A. Monsoriu, Spatial soliton formation in photonic crystal fibers, Opt. Exp. **11**, 452 (2003).

[65] A. Ferrando, M. Zacarés, P. Fernández de Córdoba, D. Binosi, and J. A. Monsoriu, *Vortex solitons in photonic crystal fibers*, Opt. Exp. **12**, 817 (2004).

[66] J. W. Fleischer, M. Segev, N. K. Efremidis, and D. N. Christodoulides, *Observation of two-dimensional discrete solitons in optically induced nonlinear photonic lattices*, Nature **422**, 147 (2003).

[67] J. W. Fleischer, G. Bartal, O. Cohen, O. Manela, M. Segev, J. Hudock, and D. N. Christodoulides, *Observation of vortex-ring "discrete" solitons in 2D photonic lattices*, Phys. Rev. Lett. **92**, 123904 (2004).

[68] K. Fradkin-Kashi, A. Arie, P. Urenski, and G. Rosenman, *Multiple nonlinear optical interactions with arbitrary wave vector differences*, Phys. Rev. Lett. **88**, 023903 (2002).

[69] I. R. Gabitov and S. K. Turitsyn, *Averaged pulse dynamics in a cascaded transmission system with passive dispersion compensation*, Opt. Lett. **21**, 327 (1996).

[70] J. J. García-Ripoll and V. M. Pérez-García, *Barrier resonances in Bose-Einstein condensation*, Phys. Rev. A **59**, 2220 (1999).

[71] J. J. García-Ripoll,V. M. Pérez-García, and P. Torres, *Extended parametric resonances in nonlinear Schrödinger systems*, Phys. Rev. Lett. **83**, 1715 (1999).

[72] C. S. Gardner, J. M. Green, M. D. Kruskal, and R. M. Miura, *Method for solving the Korteweg De Vries equation*, Phys. Rev. Lett. **19**, 1095 (1967).

[73] J. Garnier, *Stabilization of dispersion-managed solitons in random optical fibers by strong dispersion management*, Opt. Commun. **206**, 411 (2002).

[74] B. V. Gisin and A. A. Hardy, *Stationary solutions of plane nonlinear optical antiwaveguides*, Opt. Quant. Electr. **27**, 565 (1995).

[75] B. V. Gisin, A. Kaplan, and B. A. Malomed, *Spontaneous symmetry breaking and switching in planar nonlinear antiwaveguide*, Phys. Rev. E **61**, 2804 (2000).

[76] R. Grimshaw, J. He, and B.A. Malomed, *Decay of a soliton in a periodically modulated nonlinear waveguide*, Phys. Scripta **53**, 385 (1996).

[77] A. Gubeskys, B. A. Malomed, and I. M. Merhasin, *Alternate solitons: Nonlinearly-managed one- and two-dimensional solitons in optical lattices*, Stud. Appl. Math. **115**, 255 (2005).

[78] M. Gutin, U. Mahlab, and B. A. Malomed, *Shaping NRZ pulses and suppression of the inter-symbol interference by a second-harmonic-generating module*, Opt. Commun. **200**, 401 (2001).

[79] A. Hasegawa and F. Tappert, *Transmission of stationary nonlinear optical pulses in dispersive dielectric fibers. I. Anomalous dispersion*, Appl. Phys. Lett. **23**, 142 (1973); *Transmission of stationary nonlinear optical pulses in dispersive dielectric fibers. II. Normal dispersion, ibid.* **23**, 171 (1973).

[80] R. Hasse, *Schrödinger solitons and kinks behave like Newtonian particles*, Phys. Rev. A **25**, 583 (1982).

[81] S. Inouye, M. R. Andrews, J. Stenger, H.-J. Miesner, D. M. Stamper-Kurn, and W. Ketterle, *Observation of Feshbach resonances in a Bose–Einstein condensate*, Nature **392**, 151 (1998).

[82] Y. Kagan, E. L. Surkov, and G. V. Shlyapnikov, *Evolution and global collapse of trapped Bose condensates under variations of the scattering length*, Phys. Rev. Lett. **79**, 2604 (1997).

[83] A. A. Kanashov and A. M. Rubenchik, *On diffraction and dispersion effect on three wave interaction*, Physica D **4**, 122 (1981).

[84] T. Kapitula, P. G. Kevrekidis, and B. A. Malomed, *Stability of multiple pulses in discrete systems*, Phys. Rev. E 63, 036604.

[85] A. Kaplan, B. V. Gisin, and B. A. Malomed, *Stable propagation and all-optical switching in planar waveguide-antiwaveguide periodic structures*, J. Opt. Soc. Am. B 19, 522 (2002).

[86] Yu. N. Karamzin and A. P. Sukhorukov, JETP Lett. 11, 339 (1974).

[87] D. J. Kaup, B. A. Malomed, and R. S. Tasgal, *Internal dynamics of a vector soliton*, Phys. Rev. E 48, 3049 (1993).

[88] D. J. Kaup, B. A. Malomed, and J. Yang, *Collision-induced pulse timing jitter in a wavelength-division-multiplexing system with strong dispersion management*, J. Opt. Soc. Am. B 16, 1628 (1999).

[89] P. G. Kevrekidis, B. A. Malomed, and A. R. Bishop, *Bound states of two-dimensional solitons in the discrete nonlinear Schrödinger equation*, J. Phys. A Math. Gen. 34, 9615 (2001).

[90] P. G. Kevrekidis, G. Theocharis, D. J. Frantzeskakis, and B. A. Malomed, *Feshbach resonance management for Bose-Einstein condensates*, Phys. Rev. Lett. 90, 230401 (2003).

[91] Y. S. Kivshar and B. Luther-Davies, *Dark optical solitons: physics and applications*, Phys. Rep. 298, 81 (1998).

[92] L. Khaykovich, F. Schreck, G. Ferrari, T. Bourdel, J. Cubizolles, L. D. Carr, Y. Castin, and C. Salomon, *Formation of a matter-wave bright soliton*, Science 296, 1290-1293 (2002).

[93] Yu. S. Kivshar and B. A. Malomed, *Dynamics of solitons in nearly integrable systems*, Rev. Mod. Phys. 61, 763-911 (1989).

[94] J. C. Knight, T. A. Birks, P. St. J. Russell and D. M. Atkin, *All-silica single-mode optical fiber with photonic crystal cladding*, Opt. Lett. 21, 1547 (1996).

[95] F. M. Knox, W. Forysiak, and N. J. Doran, *10-Gbt/s soliton communication systems over standard fiber at 1.55 μm and the use of dispersion compensation*, IEEE J. Lightwave Tech. 13, 1955 (1995).

[96] V. V. Konotop and L. Vázquez. *Nonlinear Random Waves* (World Scientific: Singapore, 1994).

[97] K. Hayata and M. Koshiba, *Multidimensional solitons in quadratic nonlinear media*, Phys. Rev. Lett. 71, 3275 (1993).

[98] D. Krökel, N. J. Halas, G. Giuliani, and D. Grischkowsky, *Dark-pulse propagation in optical fibers*, Phys. Rev. Lett. 60, 29-32 (1988).

[99] T. I. Lakoba and D. J. Kaup, *Hermite-Gaussian expansion for pulse propagation in strongly dispersion managed fibers*, Phys. Rev. E **58**, 6728 (1998).

[100] T. Lakoba, J. Yang, D. J. Kaup, and B. A. Malomed, *Conditions for stationary pulse propagation in the strong dispersion management regime*, Opt. Commun. **149**, 366 (1998).

[101] X. Liu, K. Beckwitt, and F. Wise, *Two-dimensional optical spatiotemporal solitons in quadratic media*,

Phys. Rev. E **62**, 1328 (2000).

[102] X. Liu, L. Qian, and F. Wise, *High-energy pulse compression by use of negative phase shifts produced by the cascaded* $\chi^{(2)} : \chi^{(2)}$ *nonlinearity*, Opt. Lett. **24**, 1777 (1999).

[103] X. Liu, L. J. Qian, F. M. Wise, *Generation of optical spatiotemporal solitons*, Phys. Rev. Lett. **82**, 4631 (1999).

[104] B. A. Malomed, *Variational methods in nonlinear fiber optics and related fields*, Progress in Optics **43**, 69-191 (E. Wolf, editor: North Holland, Amsterdam, 2002).

[105] B. A. Malomed, *Nonlinear Schrödinger equations*, in: Encyclopedia of Nonlinear Science, pp. 639-643, ed. by A. Scott (New York: Routledge, 2005).

[106] B. A. Malomed and A. Berntson, *Propagation of an optical pulse in a fiber link with random dispersion management*, J. Opt. Soc. Am. B **18**, 1243 (2001).

[107] B. A. Malomed, P. Drummond, H. He, A. Berntson, D. Anderson, and M. Lisak, *Spatio-temporal solitons in optical media with a quadratic nonlinearity*, Phys. Rev. E **56**, 4725 (1997).

[108] B. A. Malomed, T. Mayteevarunyoo, E. A. Ostrovskaya, and Y. S. Kivshar, *Coupled-mode theory for spatial gap solitons in optically-induced lattices*, Phys. Rev., in press (2005).

[109] B. A. Malomed, D. Mihalache, F. Wise, and L. Torner, *Spatiotemporal optical solitons*, J. Opt. B: Quant. Semicl. Opt., **7**, R53 (2005).

[110] B. A. Malomed, D. F. Parker, and N. F. Smyth, *Resonant shape oscillations and decay of a soliton in periodically inhomogeneous nonlinear optical fiber*, Phys. Rev. E **48**, 1418 (1993).

[111] B.A. Malomed, G. D. Peng, P. L. Chu, I. Towers, A. V. Buryak, and R. A. Sammut, *Stable helical solitons in optical media*, Pramana **57**, 1061 - 1078 (2001).

[112] B. A. Malomed and N. F. Smyth, *Resonant splitting of a vector soliton in a periodically inhomogeneous birefringent optical fiber*, Phys. Rev. E 50, No. 2, 1535-1542 (1994).

[113] B. A. Malomed and R. S. Tasgal, *Vibration modes of a gap soliton in a nonlinear optical medium*, Phys. Rev. E **49**, 5787 (1994).

[114] B. A. Malomed and R. S. Tasgal, *Internal vibrations of a vector soliton in coupled nonlinear Schrödinger equations*, Phys. Rev. E **58**, 2564 (1998).

[115] B. A. Malomed, Z. H. Wang, P. L. Chu, and G. D. Peng, *Multichannel switchable system for spatial solitons*, J. Opt. Soc. Am. B **16**, 1197 (1999).

[116] P. V. Mamyshev and N. A. Mamysheva, *Pulse-overlapped dispersion-managed data transmission and intrachannel four-wave mixing*, Opt. Lett. **24**, 1454 (1999).

[117] S. V. Manakov, *On the theory of two-dimensional stationary self focussing of electromagnetic waves*, Zh. Eksp. Teor. Fiz. **65**, 505-516 (1973) [in Russian; English translation: Sov. Phys. JETP **38**, 248-253 (1974)].

[118] S. Maneuf and F. Reynaud, *Quasi-steady state self-trapping of first, second, and third order subnanosecond soliton beams*, Opt. Commun. **66**, 325 (1988).

[119] A. Matijosius, J. Trull, P. Di Trapani, A. Dubietis, R. Piskarskas, A. Varanavicius, and A. Piskarskas, *Nonlinear space-time dynamics of ultrashort wave packets in water*, Opt. Lett. **29**, 1123 (2004).

[120] M. Matsumoto, *Theory of stretched-pulse transmission in dispersion-managed fibers*, Opt. Lett. **22**, 1238 (1997).

[121] M. Matsumoto, *Instability of dispersion-managed solitons in a system with filtering*, Opt. Lett. **23**, 1901 (1998).

[122] M. Matuszewski, M. Trippenbach, B. A. Malomed, E. Infeld, and A. A. Skorupski, *Two-dimensional dispersion-managed light bullets in Kerr media*, Phys. Rev. E **70**, 016603 (2004).

[123] M. Matuszewski, E. Infeld, B. A. Malomed, and M. Trippenbach, *Stabilization of three–dimensional light bullets by a transverse lattice in a Kerr medium with dispersion management*, Opt. Commun., in press (2005).

[124] G. McConnell and E. Riis, *Ultra-short pulse compression using photonic crystal fibre*, Appl. Phys. B: Lasers Opt. 78, 557 (2004).

[125] D. Mihalache, D. Mazilu, F. Lederer, Y. V. Kartashov, L.-C. Crasovan, and L. Torner, **Stable three-dimensional spatiotemporal solitons in a two-dimensional photonic lattice**, Phys. Rev. E **70**, 055603(R) (2004).

[126] J. McEntee, *Solitons go the distance in ultralong-haul DWDM*, Fibre Systems Europe, January 2003, p. 19.

[127] L. F. Mollenauer, R. H. Stolen, and J. P. Gordon, *Experimental observation of picosecond pulse narrowing and solitons in optical fibers*, Phys. Rev. Lett. **45**, 1095 (1980).

[128] G. D. Montesinos, V. M. Perez-Garcia, and H. Michinel, *Stabilized two-dimensional vector solitons*, Phys. Rev. Lett. **92**, 133901 (2004).

[129] G. D. Montesinos and V. M. Pérez-García, H. Michinel, and J. R. Salgueiro, *Stabilized vortices in layered Kerr media*, Phys. Rev. E **71**, 036624 (2005).

[130] G. D. Montesinos, V. M. Perez-Garcia, and P. J. Torres, *Stabilization of solitons of the multidimensional nonlinear Schrödinger equation: matter-wave breathers*, Physica D **191**, 193 (2004).

[131] M. Nakazawa and H. Kubota, *Optical soliton communication in a positively and negatively dispersion-allocated optical-fiber transmission-line*, Electr. Lett. **31**, 216 (1995).

[132] D. N. Neshev, T. J. Alexander, E. A. Ostrovskaya, and Y. S. Kivshar, H. Martin, I. Makasyuk, and Z. Chen, *Observation of Discrete Vortex Solitons in Optically Induced Photonic Lattices*, Phys. Rev. Lett. **92**, 123903 (2004).

[133] A. C. Newell. *Solitons in Mathematics and Physics* (SIAM: Philadelphia, 1985).

[134] A. N. Niculae, W. Forysiak, A. J. Gloag, J. H. B. Nijhof, and N. J. Doran, *Soliton collisions with wavelength-division multiplexed systems with strong dispersion management*, Opt. Lett. **23**, 1354 (1998).

[135] E. A. Ostrovskaya and Y. S. Kivshar, *Matter-wave gap solitons in atomic band-gap structures*, Phys. Rev. Lett. **90**, 160407 (2003).

[136] E. A. Ostrovskaya and Y. S. Kivshar, *Matter-Wave Gap Vortices in Optical Lattices*, Phys. Rev. Lett. **93**, 160405 (2004).

[137] C. Paré, and P.-A. Bélanger, Antisymmetric soliton in a dispersion-managed system, Opt. Commun. **168**, 103 (1999).

[138] C. Paré, V. Roy, F. Lesage, P. Mathieu, and P.-A. Bélanger, *Coupled-field description of zero-average dispersion management*, Phys. Rev. E **60**, 4836 (1999).

[139] C. Paré, A. Villeneuve, P.-A. Belangé, and N.J. Doran, *Compensating for dispersion and the nonlinear Kerr effect without phase conjugation*, Opt. Lett. **21**, 459 (1996).

[140] D. E. Pelinovsky, *Instabilities of dispersion-managed solitons in the normal dispersion regime*, Phys. Rev. E **62**, 4283 (2000).

[141] C. J. Pethik and H. Smith. *Bose-Einstein Condensation in Dilute Gases*. (Cambridge University Press: Cambridge, 2002).

[142] V. M. Pérez-García, H. Michinel, J. I. Cirac, M. Lewenstein, and P. Zoller, *Low energy excitations of a Bose-Einstein condensate: A time-dependent variational analysis*, Phys. Rev. Lett. **77**, 5320 (1996).

[143] V. M. Pérez-García, H. Michinel, J. I. Cirac, M. Lewenstein, and P. Zoller, *Dynamics of Bose-Einstein condensates: Variational solutions of the Gross-Pitaevskii equations*, Phys. Rev. A **56**, 1424 (1997).

[144] V. M. Pérez-García, P. Torres, J. J. Garcia-Ripoll, and H. Michinel, *Moment analysis of paraxial propagation in a nonlinear graded index fibre*, J. Opt. B: Quantum Semiclass. Opt. **2**, 353 (2000).

[145] L. M. Pismen. *Vortices in Nonlinear Fields* (Oxford University Press: Oxford, 1999).

[146] E. Poutrina and G. P. Agrawal, *Design rules for dispersion-managed soliton systems*, Opt. Commun. **206**, 193 (2002).

[147] R. H. Rand, *Lecture Notes on Nonlinear Vibrations* (http://www.tam.cornell.edu/randdocs/nlvibe45.pdf).

[148] J. L. Roberts, N. R. Claussen, J. P. Burke, Jr., C. H. Greene, E. A. Cornell, and C. E. Wieman, *Resonant Magnetic Field Control of Elastic Scattering in Cold ^{85}Rb*, Phys. Rev. Lett. **81**, 5109 (1998).

[149] J. S. Russell, *Report on Waves*, Rep. 14th Meeting British Assoc. Adv. Sci., p. 311 (1844).

[150] H. Saito and M. Ueda, *Dynamically stabilized bright solitons in a two-dimensional Bose-Einstein condensate*, Phys. Rev. Lett. **90**, 040403 (2003).

[151] H. Sakaguchi and B.A. Malomed, *Dynamics of positive- and negative-mass solitons in optical lattices and inverted traps*, J. Phys. B **37**, 1443 (2004).

[152] H. Sakaguchi and B. A. Malomed, *Two-dimensional loosely and tightly bound solitons in optical lattices and inverted traps*, J. Phys. B: At., Mol. Opt. Phys. **37**, 2225 (2004).

[153] H. Sakaguchi and B. A. Malomed, *Resonant nonlinearity management for nonlinear Schrödinger solitons*, Phys. Rev. E **70**, 066613 (2004).

[154] J. Satsuma and N. Yajima N, *Initial value problems of one-dimensional self-modulation of nonlinear waves in dispersive media*, Progr. Theor. Phys. Suppl. No. 55, pp.284-306 (1974).

[155] R. Schiek, Y. Baek, and G. I. Stegeman, *One-dimensional spatial solitary waves due to cascaded second-order nonlinearities in planar waveguides*, Phys. Rev. E **53**, 1138 (1996).

[156] Y. Silberberg, *Collapse of Optical Pulses*, Opt. Lett. **15**, 1282 (1990).

[157] J. Stenger, S. Inouye, M. R. Andrews, H. J. Miesner, D. M. Stamper-Kurn, and W. Ketterle, *Strongly enhanced inelastic colllisions in a Bose-Einstein condensate near Feshbach resonances*, Phys. Rev. Lett. **82**, 2422 (1999).

[158] K. E. Strecker, G. B. Partridge, A. G. Truscott, and F. G. Hulet, *Formation and propagation of matter-wave soliton trains*, Nature **417**, 150-153 (2002).

[159] C. Sulem and P.-L. Sulem. *The Nonlinear Schrödinger Equation* (Springer: New York, 1999).

[160] M. Suzuki, I. Morita, N. Edagawa, S. Yamamoto, H. Taga, and S. Akiba, *Reduction of Gordon-Haus timing jitter by periodic dispersion compensation in soliton transmission*, Electron. Lett. **31**, 2027 (1995).

[161] H. Toda, Y. Furukawa, T. Kinoshita, Y. Kodama, and A. Hasegawa, *Optical soliton transmission experiment in a comb-like dispersion profiled fiber loop*, IEEE Phot. Tech. Lett. **9**, 1415 (1997).

[162] L. Torner, *Walkoff-compensated dispersion-mapped quadratic solitons*, IEEE Phot. Tech. Lett. **11**, 1268 (1999).

[163] L. Torner, S. Carrasco, J. P. Torres, L. C. Crasovan, and D. Mihalache, *Tandem light bullets*, Opt. Commun. **199**, 277 (2001).

[164] W. E. Torruellas, Z. Wang, D. J. Hagan, E. W. VanStryland, G. I. Stegeman, L. Torner, and C. R. Menyuk, *Observation of Two-Dimensional Spatial Solitary Waves in a Quadratic Medium*, Phys. Rev. Lett. **74**, 5036 (1995).

[165] I. Towers and B. A. Malomed, *Stable (2+1)-dimensional solitons in a layered medium with sign-alternating Kerr nonlinearity*, J. Opt. Soc. Am. B **19**, 537 (2002).

[166] M. Trippenbach, M. Matuszewski, and B. A. Malomed, *Stabilization of three-dimensional matter-waves solitons in an optical lattice*, Europhys. Lett. **70**, 8 (2005).

[167] S. K. Turitsyn and Mezentsev, *Dynamics of self-similar dispersion-managed soliton presented in the basis of chirped Gauss-Hermite functions*, JETP Lett. **67**, 640-646 (1998).

[168] S. K. Turitsyn, E. G. Shapiro, S. B. Medvedev, M. P. Fedoruk, and V. K. Mezentsev, *Physics and mathematics of dispersion-managed optical solitons*, Compt. Rend. Phys. **4**, 145 (2003).

[169] T. Ueda and W. L. Kath, *Dynamics of coupled solitons in nonlinear optical fibers*, Phys. Rev. A **42**, 563 (1990).

[170] A. V. Ustinov, *Solitons in Josephson junctions*, Physica D **123**, 315 (1998).

[171] M. Wald, B. A. Malomed, and F. Lederer, *Interaction of moderately dispersion-managed solitons*, Opt. Commun. **172**, 31 (1999).

[172] A. M. Weiner, J. P. Heritage, R. J. Hawkins, R. N. Thurston, E. M. Kirschner, D. E. Leaird, and W. J. Tomlinson, *Experimental observation of the fundamental dark soliton in optical fibers*, Phys. Rev. Lett. **61**, 2445 (1988).

[173] P. Xie, Z.-Q. Zhang, and X. Zhang, *Gap solitons and soliton trains in finite-sized two-dimensional periodic and quasiperiodic photonic crystals*, Phys. Rev. E **67**, 026607 (2003).

[174] J. Yang and Z. H. Musslimani, *Fundamental and vortex solitons in a two-dimensional optical lattice*, Opt. Lett. **28**, 2094 (2003).

[175] N. Zabusky and M. Kruskal, *Interactions of solitons in a collisionless plasma and the recurrence of initial states*, Phys. Rev. Lett. **15**, 240-243 (1965).

[176] V. E. Zakharov, S. V. Manakov, S. P. Novikov, and L. P. Pitaevskii. *Solitons: the Inverse Scattering Transform Method* (Nauka publishers: Moscow, 1980 (in Russian)) [English translation: Consultants Bureau, New York, 1984].

[177] V. E. Zakharov and A. B. Shabat, *Exact theory of two-dimensional self-focusing and one-dimensional self-modulation of waves in nonlinear media*, Zh. Eksp. Teor. Fiz. **61**, 118 (1971) [English translation: Sov. Phys. JETP **37**, 823-828 (1972)].

[178] M. Zitelli, B. Malomed, F. Matera, and M. Settembre, *Strong time jitter reduction using solitons in hyperbolic dispersion managed links*, Opt. Commun. **154**, 273 (1998).

Index